A CHEMIST'S PERSPECTIVE ON THE SHROUD OF TURIN

Raymond N. Rogers
Fellow
University of California
Los Alamos National Laboratory

Editor
Barrie M. Schwortz
Florissant, CO, USA

Cataloging-in-Publication Data
Rogers, Raymond N.
A Chemist's Perspective on the Shroud of Turin
ISBN 978-0-6152-3928-6
Publisher: Barrie M. Schwortz
www.shroud.com
July 2008

TABLE OF CONTENTS

Chapter	Page

As a practicing scientist, it came as a shock to get involved with a project that stirred such emotions as the Shroud of Turin. It should be possible to use Scientific Method on anything that can be observed, but deep belief systems can make science difficult. Science is both a body of knowledge and a way of solving problems. We all use science in everyday living. This offering is an attempt to allow nonscientists to view the object through the eyes of an obligate scientist.

When you are handed a real thing, something you can see and feel but have never seen before, your mind enters into a natural process called "whazzat?" [Translation: what-is-that?] You look for sharp edges (is it a cutting tool?), and you look for a handle (do I use it or does it fit some appliance?). You try to observe every feature of the whazzat? you can see or feel. You are using a natural, human mental process you have learned by experience. You can call it "Scientific Method."

Fortunately for chemists, the Shroud was severely burned during the fire of 1532. The fire provided an excellent "chemical test" on all of the materials of the Shroud. Before it was possible to make direct observations on the cloth, we were able to observe the effects of the fire on the image, apparent blood stains, and clear parts of the cloth by looking at high-resolution photographs. The chemistry of the fire is still important.

The image is the result of many superficial yellow linen fibers. The color is discontinuously distributed in image areas, and it does not penetrate either the cloth or the individual, microscopic linen fibers. Image distribution poses a perplexing problem for scientists.

The image is extremely faint, and its details could not be observed until the advent of photography. Observations by contrast-enhanced and ultraviolet-fluorescence photography have been extremely important in attempts to understand the Shroud.

Samples are difficult to obtain for scientific observations. A small segment of cloth was cut for textile analysis in 1973, and I was allowed to take samples in 1978 with special adhesive tape. I was allowed to make observations on two, well-documented, small yarn segments from the 1988 radiocarbon sample in 2003. Observations on the samples have been invaluable, and a few photomicrographs are shown. See Chapter IX for details on the radiocarbon sample.

Colored image fibers appear only on the very topmost parts of the weave, and significant parts of the weave in an image area are not colored at all between the spots of color. The color does not penetrate the cloth to a significant depth in most places; however, photographs of the back surface of the cloth that were taken in June and July of 2002 seem to indicate that light image color appears in the area of the hair. Whatever mechanism produced the image color did not significantly penetrate the cloth above areas of skin but might have above the hair.

The primary assignment given the chemical section of STURP was to investigate the claim that the image had been painted. We decided that both chemical tests and spectrometry would be required to confirm specific pigments; however, we got many surprises. Image color does not appear under the blood stains, and it appears only on the surfaces of image fibers.. This indicates mild conditions for image formation.

The Shroud was observed by visible and ultraviolet spectrometry, infrared spectrometry, x-ray fluorescence spectrometry, and thermography. Later observations were made by pyrolysis-mass-spectrometry and laser-microprobe Raman analyses.

In 1988, the custodians allowed samples to be cut for radiocarbon age determination. Three laboratories analyzed aliquots, and they reported that "The age of the shroud is obtained as AD 1260-1390, with at least 95% confidence." Comparisons between image and non-image Shroud fibers and the Raes and radiocarbon samples showed that there are large chemical differences between the main part of the cloth and the area from which the radiocarbon sample was cut. The radiocarbon age determination was run on an invalid sample, and the age reported is incorrect for the time the flax was cut to make the linen.

EDITOR'S NOTE

When Ray Rogers asked me to edit this book in 2003, I was deeply honored and welcomed the opportunity to help my good friend with his important work. Ray had often relied on me as his "diplomatic" editor on other Shroud materials he had written and frequently reminded me this was necessary because he himself was no diplomat. (In the world of the Shroud, diplomacy is truly an issue).

We worked on this book for over a year and I carefully went through it with him, chapter by chapter, to make sure everything was clear and it included no obvious errors. I was barely qualified to read, let alone edit any of his scientific material, so I simply checked it for typos or any other obvious mistakes.

Work on the book was set aside as Ray took up what would become his final research on the Shroud, and which culminated in the publication of his Thermochimica Acta paper in January 2005. About a year before he died, in a telephone conversation we had in which we discussed the upcoming publication of that paper, I suggested that there would soon be serious media interest in his work and he would inevitably be asked for interviews.

Ray was very quick to remind me that he did not give interviews, period. However, to my sincere amazement, he invited me to come to his home with my video equipment, sleep in his spare bedroom and promised he would spend as much time as his failing health allowed sitting in front of my camera, talking to me about the Shroud. I immediately arranged to take him up on his offer, and ultimately spent five days and nights with Ray and Joan in their home in Los Alamos, New Mexico, in May, 2004.

One evening during my stay, the subject of this book came up again. As we reviewed it in his laptop computer, I noticed that he had included my name under his on the title page. I told him that I truly appreciated the compliment but strongly felt my name should not be there at all, as I was simply the copy editor and had contributed virtually nothing to the content of the book. He explained that he had put it there for a reason. It was there because he wanted *me* to make sure this book got published. In fact, he made me *promise* him that I would do so and added the caveat that I had to insure that no book editor would "mess" with his science! I gave him my solemn word that I would do so, and you now hold the result of that promise in your hands.

Ray died about ten months after we had that conversation and Joan Rogers felt we should wait a while before starting work on the book again. Both of us were mourning his loss and frankly, neither of us really had the motivation for the task. But three years have now gone by since his death and this year marks the 30th anniversary of STURP's historic 1978 examination of the Shroud, in which Ray played such an important part. We both agreed that this was the perfect time to fulfill my promise and give everyone the opportunity to read Ray's words for themselves.

Barrie Schwortz - July 5, 2008

PREFACE

I have supported myself as a chemist for 52 years, and I have studied archaeology, anthropology, geochronology, and soil chemistry at the graduate level. I have done chemical analyses of cultural materials as a favor to my archaeologist friends since 1948, and I took a sabbatical to do research on organic materials in ancient-man sites. As a result of my interest and publications in archaeology, I was asked to take part in a "scientific study" on the shroud of Turin in 1977. This became known as the Shroud of Turin Research Project (STURP).

I had never heard of the relic. I had no interest in religious objects of devotion. I probably should have rejected any involvement in the Shroud of Turin project, but some curious observations appeared on photographs of the cloth. I thought it would be a simple matter to identify paint, and I told a colleague at work, "Give me Classical Scientific Method and 20 minutes and I'll have that thing shot full of holes." I entered the project without any preconceptions, and I believe that one of my best contributions has been objectivity. But objectivity is not always greeted with joy. Perhaps three death threats from devout Christians have colored my outlook, but they have nearly been balanced by violently antireligious "scientists." If both fringes hate you, you must be doing something right. I dislike persons or groups who demand that all published evidence support their narrow view of truth. The first fruit of my involvement in the project was a personal recognition that there really are two fanatical fringes on the bell-shaped curve of humanity.

I quickly found that the Shroud was not a simple puzzle to solve. That is what has made it so interesting, but it can be studied according to Classical Scientific Method. Now, in 2003, after nearly 26 years of contemplation, study, and analysis, some scientific facts seem to be more clearly visible.

Many persons truly expected the "scientific" study of the Shroud to prove the resurrection of Jesus. Several scientists, who should have maintained their objectivity, selected evidence to make it appear that the Shroud was a forgery or hoax. I can not present a comprehensive analysis of all of the claims about the Shroud; however, I hope to supply as comprehensive a discussion as possible on the tools each reader can use to make personal tests that are based on facts. My discussions will be biased toward chemical evidence; however, I think that is a valid bias. The Shroud is made out of chemical compounds, and all of its visible features are the result of other compounds or chemical reactions.

Science is both a body of knowledge and a way of solving problems. We all use science in everyday living. We accumulate knowledge of the real world, and it helps keep us alive among the hazards of daily living. When we are faced with a problem, we make assumptions on the best course, and we check our assumptions against what we know to be real. If we choose to believe that driving a speeding car over a cliff won't hurt us, we lose. The Roman Church could make persons profess its dogma during the inquisition. That did not keep the earth from rotating around the sun. Reality and truth will always win in the end. Both the Church and science need a good dose of objectivity and reality.

Science is very good at *proving* that some things are *impossible*. You can not make a perpetual-motion machine. However, science can not deliver absolute truths from Olympus. Statements that "this is the *only way*" something could happen are not "scientific." When a scientist faces a problem, he or she will try to find out what other people have discovered about it and will then formulate a *hypothesis* about the problem. The hypothesis will then be *tested* against all of the observations that can be assembled. No pertinent observations can be thrown out just because they don't fit. Something approaching the truth will be found to include all of the confirmed, real observations. That is how science works. Science is deductive: Religions are revealed.

Some observations are too tenuous to consider. When a person says, "I think I see," the evidence is probably weak. People claim to see flowers and writing on the cloth, but none of these observations have been confirmed by independent analysis. I will not report or use such evidence in this presentation.

When any hypothesis looks pretty strong, it will be used to make predictions, and the predictions will be tested experimentally. If it appears to be totally reliable within its range of testing, the hypothesis may be advanced to the higher status of a scientific *theory*. After a theory has survived all of the testing that a few generations of scientists can think up, it may be called a *law*. No answers to the most perplexing questions about the Shroud can yet be called a theory.

Einstein published his Theory of General Relativity in 1915. It could be called a theory, because it was developed from "first principles" by rigorous mathematical derivation. Einstein's theories have been used to make successful predictions in cosmology and nuclear physics, but the testing is still going on. Some aspects of his theories have been extremely difficult to test by experiments. Some aspects of Shroud observations seem to be extremely difficult to test: ***That does not mean that miracles were involved.*** Persons who want to prove miracles tend to stop testing before they have to reject their favorite "theory."

I believe that some observations can, in the best scientific tradition, eliminate some hypotheses about the Shroud. For example, energetic radiation or high temperatures could not have caused the image. These are favorite "theories," because they can be used to prove miraculous events at the time of the Biblical resurrection. I will attempt to provide as much evidence as I can that you can use to make your own, independent tests of all hypotheses you will hear or read about the Shroud. Remember - - - all confirmed, applicable scientific observations must fit the final, true description of the Shroud's reality.

CHAPTER I: REAL THINGS

As a full-time American husband, I am often faced with that embarrassing male problem: "What is that thing"? My wife can easily blind-side me by handing me some peculiar-looking thing-a-ma-bob and saying, "Honey, what is this thing." My job is to appear completely masculine, omniscient, and unruffled while my brain comes to a boil testing all of the possible ideas I can generate instantly so that I can calmly reply, "Why Dear, that is a left-handed framistat." When there is love between two people, it is a fun game: When arrogance, hate or a desire for retribution enters into the game, it can become a nasty experience. Science is like that.

When you are handed a real thing, something you can see and feel but have never seen before, your mind enters into a natural process called "whazzat?" You look for sharp edges (is it a cutting tool?), and you look for a handle (do I use it or does it fit some appliance?). You try to observe every feature of the whazzat? you can see or feel. You are using a natural, human mental process you have learned by experience. When we call it "Scientific Method," it tends to scare you, but science is only the way you have always answered questions and solved problems. Science is based on *observation.* If you have questions about a real thing, you just naturally turn to scientific method for answers.

Sometimes a person can use all of his senses on a whazzat?, sniffing, feeling, looking, tasting, shaking and listening for a long time, maybe doing some experiments, and still not understand a "real thing." A good example often appears in archaeology. When the scientist has pondered and researched as much as he can and still can not determine a satisfactory function for some object he has found, he is likely to state from authority, "That is a ceremonial object." In seeking an understanding of the Shroud, you will often see the statement, "Science can not explain the image; therefore, it was produced by a miracle." The Greeks recognized the fallacy of such an "argumentative leap" (*non sequitur*) millennia ago. The fact that science has not *yet* found an explanation proves nothing. Science continually collects observations and information, and conclusions often change with time. Persons who believe in absolutes will be unhappy with science. Scientists will be unhappy with persons who demand that they "make up your minds." I have been called "frivolous" for modifying earlier ideas. It is just part of the job.

"Absence of evidence is not the same as evidence of absence."
Carl Sagan

The beauty in using *non sequitur* in science is that you can take it anywhere you want to go. If you want a specific "theory," all you have to do is postulate it. It also allows you to "travel light." You do not need any real facts. It would be fun to use *non sequitur* in cosmology. Let's consider the observations of "dark matter" and "dark energy."

Astronomers have observed that galaxies rotate too fast for the amount of matter they can observe in them. There must be other mass there that they can't see and does not seem to occlude the light coming from the galaxies. They also have observed that the universe has been expanding more rapidly than predicted since the Big Bang. "Science can not explain these observations": therefore, it must have a supernatural explanation. Right? It is obvious that the extra mass is the "heavenly hosts," and the expansion is a result of the angels flapping their wings in interstellar space. That being settled, we can ignore any additional observations. Isn't science wonderful?

"Science commits suicide when it adopts a creed."
Thomas Henry Huxley

Some humans take perfectly practical approaches to problem solving. The isolated tribes of New Guinea came into contact with modern technology for the first time during World War II. They would find "real things," but they would have no idea of their original function. If there was nobody around to ask, they would ponder the objects. They are members of the *Homo sapiens* species just as we are, and their minds operate in much the same fashion when solving problems. For example, they found 35-mm film cans, but they knew nothing about photography or film. Their observations showed that the cans made superb water-proof containers. Such examples of cultural "diffusion" are often observed in anthropological studies. Who are we to tell a Papuan that his use of a film can is not "true"? Devotion is a valid, true function for the Shroud, and science can not threaten that function.

Ethnology is full of accounts of meetings or collisions between different cultures. For example, Cortez marched into Tenochtitlan with gun powder and armor. The Aztecs used the tools that were available to them in an attempt to solve their problem. The tools were not up to the job. They lost their freedom, their library, and their type of culture. Cultural clashes often lead to war among humans.

Some of the bloodiest cultural clashes have occurred over belief systems. We see a clash of beliefs when we read the different accounts that have been written about the Shroud of Turin. It is a "real thing," it can be looked at, handled, and subjected to tests. You can ask questions about what makes it look the way it does, and you can test your answers to the questions. However, it is one of those rare objects that are not simply a "real thing."

Like a film can, the Shroud can have different, simultaneous functions. It can provide material for a scientific study of otherwise inaccessible history, while it provides a spiritual lift for the devout. When minds are closed, bitter disagreements result. A scientist has no right to demand that a devout person withhold devotion of the Shroud until all of the scientific evidence is in. A devout person has no right to tell a scientist that he must stop all observations, because everything that is needed is known from scriptures or feelings. Nobel scientist Richard Feynman pointed out that "it is interesting that human relationships, if there is an independent way of judging truth, can become unargumentative." Unfortunately, people often do not agree on the definition of "truth." Wars are fought and hatreds survive for millennia over such misunderstandings. The

Shroud could provide a unique opportunity to establish understanding between science and religion, but people are not up to the challenge.

The Shroud is at the center of a complex system of fundamental beliefs and emotions. Approximately 2,000 years of thoughts and writings have produced a culture we call Christianity. It did not develop the way you approach understanding "real things." It is not an insult to either science or religion to point out that there is a fundamental difference between them. Science is deduced: Religion is revealed. They are both natural parts of our lives or our mental processes.

Like the Aztecs, we can not survive as a culture or a person after we lose either our methods for solving material problems or our belief systems. Freud would say that we must have adequately balanced egos, ids, and superegos to survive as a coherent person. When we are sane, our minds are able to maintain adequate boundaries among the components of our mental processes. We can behave in a socially acceptable manner.

According to Freud, our ego functions to produce the ordinary conscious thoughts we need to direct our daily behavior; our id is mostly unconscious, instinctive thought processes; and our superego is the conscious collection of values, ideals, and taboos that set the limits for the ego and that punish us for our shortcomings by self-flagellation with guilt feelings. You might think of science as our ego and religion as our superego. When they are balanced we are "sane."

Unfortunately, in the real world we often find that the balance our mind seems to demand in order to keep us sane is not achievable. We are faced with a society that does not share our enthusiasms; that contains megalomaniacs who impose their will on us, telling us what is right and proper; and education systems that are intended to regiment rather than enlighten. When we can not balance our lives, we descend into "hysteria," which is also called conversion disorder. We can show physical symptoms that have no real organic basis. Battle fatigue is an extreme example of such a state, and it can cause paralysis, blindness, and loss of memory. When any arrogant persons or megalomaniacs assume control of our society, our culture shows similar symptoms. The arrogant and megalomaniac persons who demand their personal, absolute approach to problem solving or religion contribute to counterproductive social insanity. There should be no battle between religion and science any more than the ego or superego should control all of our actions.

The Shroud of Turin has tended to act as a wedge between science and religion. One famous scientist (a devout Episcopalian) told me, "Ray, even if you prove that thing is authentic, I won't believe it." A devout Roman Catholic told me, "I will not accept any scientific fact unless it agrees with the Gospels." Can there be any common ground?

What is this thing that causes so much emotional response? The easy answer is: "It is a 'real thing.'" Both scientists and theologians can agree on that. We diverge when we start to describe the "real thing."

3

The Shroud of Turin that we can observe with our senses is a 4.36-meter-long by 1.10-meter-wide piece of somewhat-yellowish cloth. It looks identical to museum specimens of ancient linen, except that it was woven in a 3:1 twill pattern. When you look at it carefully, you can begin to see the very faint image of a man on the cloth. The image shows both sides of the man with the front and back images of the head almost touching in the middle of the cloth. There are many burned places on the cloth as a result of a documented fire in AD 1532, and there are visible stains that look like they were made by water. The water marks are darker at the edges than they are in the middle. There are also easily visible, red stains on the cloth. They appear to have flowed into and diffused along the cloth. They are visible on the back side of the cloth. A hand-held 35-mm snapshot I took in 1978 is shown in figure I-1. It shows how faint the image is.

Monsignor Giuseppe Ghiberti published a report of a secret "restoration" of the Shroud that was done in June and July of 2002 ["Sindone le immagini 2002," Opera Diocesana Preservazione Fede - Buona Stampa, Corso Matteotti, II, 10121 Torino, Italia]. The "restoration" made major changes in the appearance and composition of the Shroud, as can be seen in full-length photographs of both the top and bottom surfaces of the cloth.

These views confirmed the fact that what appeared to be blood had seeped through the cloth. In his report, Ghiberti claimed that the only part of the image that is visible on the back surface of the cloth is the hair. The technical quality of his photographic evidence is too poor to make a scientific confirmation of that claim, and requests for access to copies of the high-resolution digital images have so far been denied. However, a faint image of the hair on the back surface does appear to be visible in Ghiberti's photographs. It is unfortunate that there was no broad-based scientific presence during these secret proceedings, because the appearance of some specific parts of the image on the back of the cloth could contribute critical information toward an understanding of the image-formation mechanism.

The largest red stain appears in the vicinity of the side and back of the image, but there are smaller stains on the head, wrist, and feet. There is a strip of cloth along one edge that appears to have been sewn to the original cloth, but it is shorter than the original. Some pieces have obviously been cut from the Shroud. There were numerous creases and folds visible before the "restoration" of 2002. One notable crease appeared directly below the jaw of the image. Many of the creases contained particulate materials. Before "restoration," thirty patches covered the worst burned places, and the whole Shroud was stitched to a backing cloth that supported the damaged material. The cloth felt flexible and strong (not at all like rotten cloth), it had a tight weave that looked almost waterproof, and the burns intersected image areas, red spots, and water stains. Some patches had been stitched onto severely scorched cloth. No unexpected odors could be detected near any part of the cloth. Aside from the burns, the Shroud appeared to be in excellent condition.

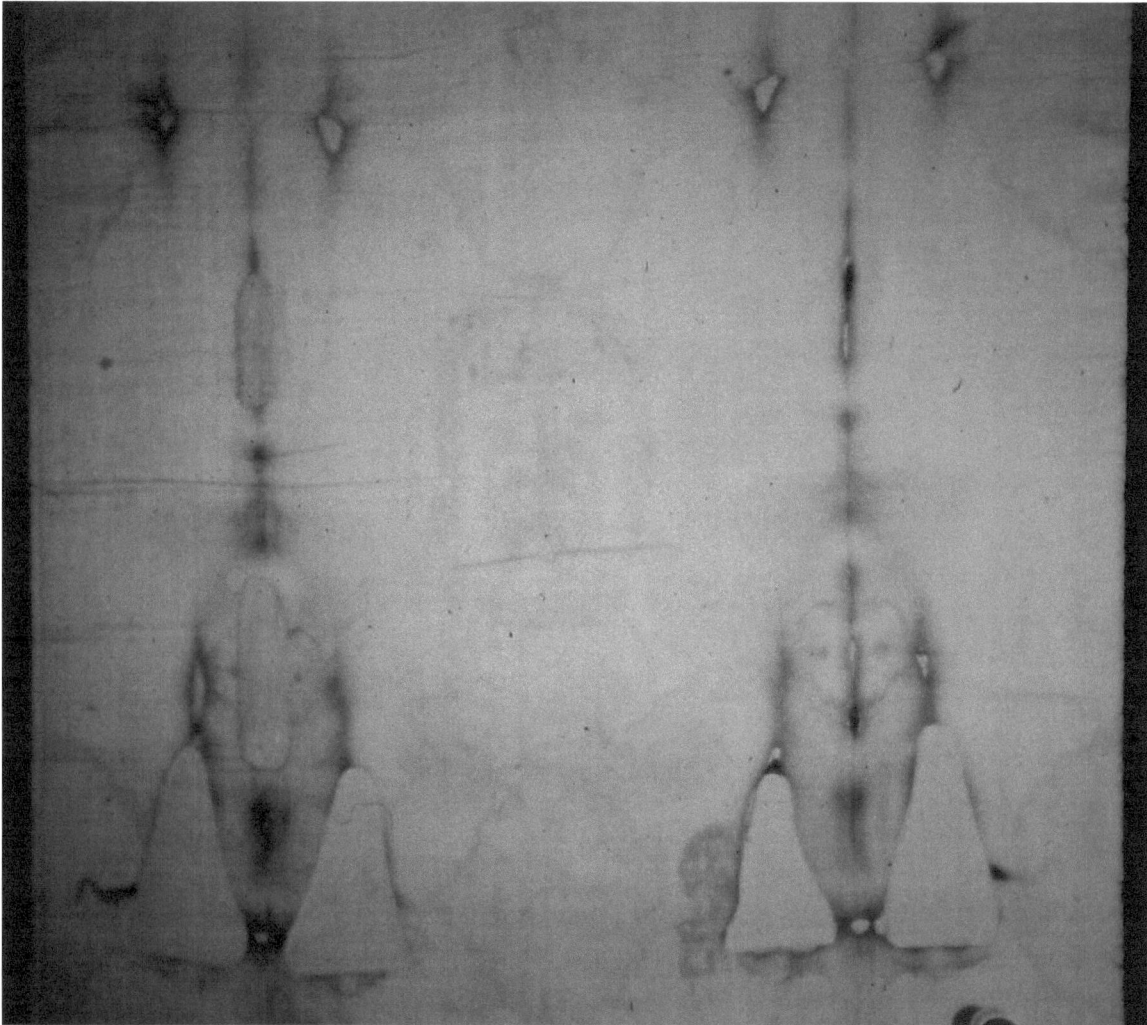

Figure I-1: A snapshot of the Shroud as I observed it in 1978. The image is very hard to see, and it is impossible to see at close range. In order to observe a specific location on the image under magnification, the observer had to be directed by another person who stood about two to three meters away from the Shroud.

During the 1978 scientific observations, the custodians of the Shroud allowed nuns to unstitch some of the seam that held the main part of the Shroud to its backing cloth so that the back side of the cloth could be observed. My Shroud research notebook for 13 October 1978 has the following notations: "Nuns unstitched the foot end of the back image around the right corner. Baima touched with his fingers. The red stain of the foot blood is visible on the back of the cloth. Do not see any sign of image on the back side - just blood. Light was very bad, and people milling around made it hard to see. Other than blood, cloth appeared to be very clear and lustrous. Nuns unstitched a little more, but still can not see image. The back side is whiter than the front. Too much of a crowd - had to leave."

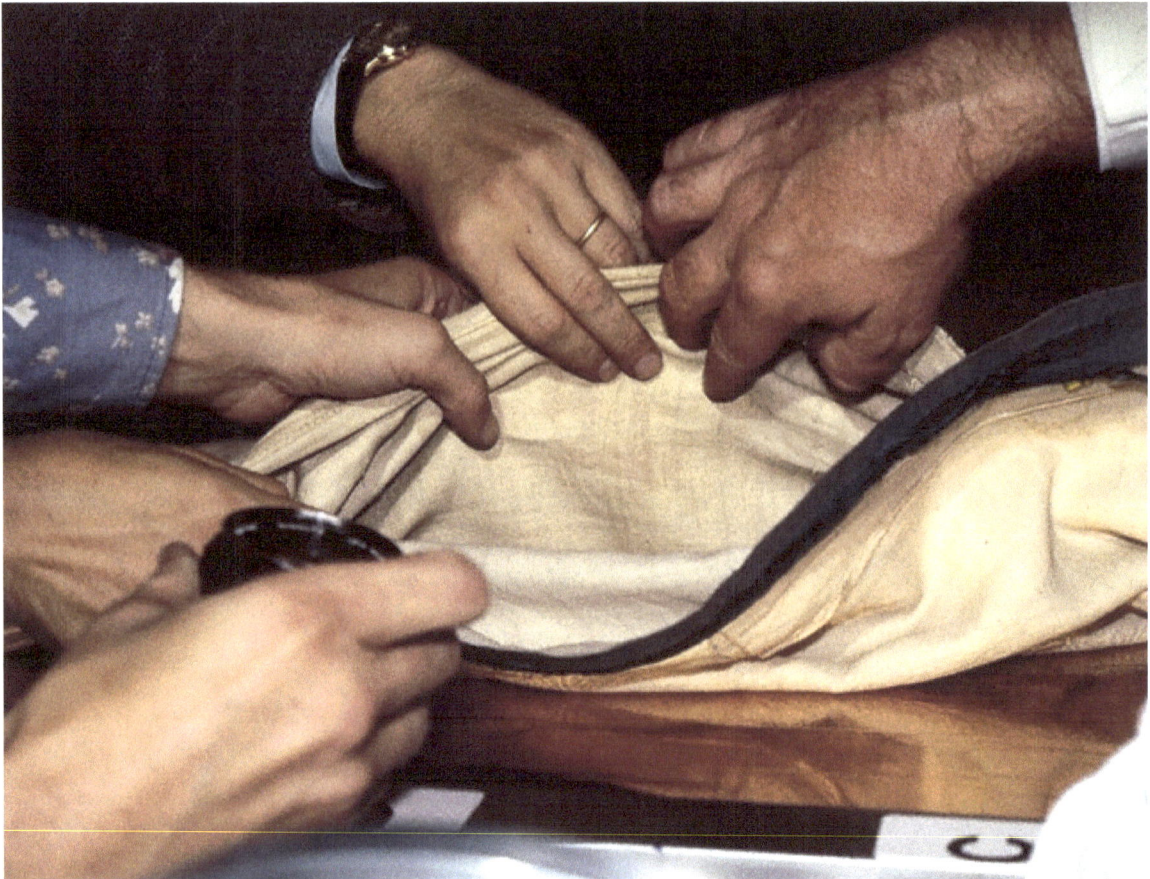

Figure I-2: In 1978, a small area of the Shroud was unstitched at the foot area of the dorsal image so the reverse side of the cloth could be examined. The red stain of the foot blood was visible on the back of the cloth, but no sign of any image was observed. ©1978 Barrie M. Schwortz

Giovanni Riggi di Numana, an Italian member of STURP, used a fiber-optics probe to look more closely at the back surface of the cloth than we could otherwise. He could observe details in the blood spots, but he could not see any trace of the image.

That group of visual observations was all that our limited senses could determine about the "real thing." Scientific methods of observation can be used to extend the reach of our senses, but the logic applied when deciding how to make observations and how to interpret scientific observations is exactly the same as we use to study any unusual whazzat?

An important part of science is to "gather all pertinent information." A scientist spends a lot of time in the library. An important part of the information on the Shroud can only be obtained from competent historians. Most "real things" have a history in time and place: where was it found, how long ago, did it look the same when it was found? Some historical facts can be documented sufficiently to be accepted in a scientific study.

The Shroud of Turin first came to public notice and was documented early in the 14th Century in France, when its custodian claimed that it was the shroud of Jesus. It has been fiercely controversial ever since. For example, the Bishop of Troyes sent a letter to the pope in AD 1389 claiming that the image had been painted and that he had interviewed the person who painted it.

The Shroud was badly damaged by fire in Chambéry, France, in 1532. It was not seen publicly again until 1534, after nuns had repaired it. It was taken by the House of Savoy to Turin, Italy, in 1578.

An international group of scientists and support personnel produced a plan for a scientific study of the Shroud in 1977, before a public exhibition in 1978. This group incorporated as the Shroud of Turin Research Project (STURP). After the 1978 public exhibition, the custodians permitted limited, restricted-access scientific studies to be made.

The custodians had allowed a small sample to be cut from the Shroud for textile analysis in 1973. Without consulting any of the scientists who had studied the cloth in 1978, the custodians allowed a much larger sample to be cut for radiocarbon dating in 1988. The date was reported as AD 1260-1390. At the time the cloth was sampled, the reliquary was treated with thymol, a phenolic compound used to sterilize materials. Thymol reacts with cellulose (linen), and the cloth's composition has undoubtedly changed.

Great care will be required to make a rigorous confirmation of the age of the cloth. As a result of the incompetent radiocarbon-sampling operation, most of the world now considers the Shroud to be a medieval hoax. Whatever the true age of the Shroud might be, an accurate age determination can not solve all of the problems associated with this peculiar relic.

During the secret "restoration" of 2002, all of the 1532 repairs were removed, the Shroud was scraped and vacuumed to remove charring, and it was stitched to a new sheet of linen. A paranoid chemist would assume that the "restoration" had been designed to eliminate any possibility of future chemical investigations. The material scraped from the scorched areas contained information on the chemical reactions of Shroud components during the fire of 1532. The surface of the Shroud might have contained the "chemical footprints" of 2,000 years of history. It doesn't now.

The charred material was largely elemental carbon. It is chemically a very unreactive material. It would not have changed significantly during the 470 years since the fire. It provided a chemical "time capsule" of Shroud history. It can easily be chemically cleaned of thymol and any microbiological deposits that might have appeared after the fire, making it an ideal material for radiocarbon dating. Before the restoration, the carbon from specific areas could have been dated separately, giving critical information about the homogeneity of the cloth as well as a "cluster" of dates. The irresponsible "restoration" after 470 years of relative peace was a disaster for Shroud science. Was it intended to be?

Theology can suggest a startling coincidental similarity between our visual observations and the written accounts of the crucifixion of Jesus of Nazareth. Many persons hoped that a scientific study of the Shroud of Turin could prove the "authenticity" of both the Shroud and the Gospel accounts. The definition of the word "authentic" depended on the person using it. It might mean that it is an authentic shroud, that it was the shroud used to wrap Jesus after the crucifixion, or that it proved the resurrection. At that point the simmering conflict between science and religion boiled. There is a limit to the information that science can accept or provide.

Feynman said: "The principle that observation is the judge imposes a severe limitation to the kind of questions that can be answered." And, "...if a thing is not scientific, if it cannot be subjected to the test of observation, this does not mean that it is dead, or wrong, or stupid." "There are some things left out, for which the method does not work. This does not mean that those things are unimportant. They are, in fact, in many ways the most important."

Historians are often faced with sorting historical facts from myths or beliefs. A physical scientist must make the same effort. What "facts of history" can be accepted for incorporation into a scientific study of the Shroud? One good example of a historical fact that can be accepted is the documentation for the fire of 1532. The date is established historically, and the effects can be observed scientifically. Claims that the Shroud protected a city or cured diseases can not be established by observation, and they can not be accepted as part of a scientific study. Many claims have been made about what can be observed on the Shroud, some have been quite fanciful. If they can not be confirmed by independent observation, they can not be accepted.

While we were testing our scientific instruments with a dry run in Connecticut, an ecstatic member of a religious group said: "I am so glad that you are going to *prove* the resurrection." That illustrates the differences between scientific goals and religious goals. A scientist or a husband faced with a "whazzat?" tests ideas of what something may be. He does *not* "declare" something and proceed to prove that something at all cost. There is a siege mentality implicit in trying to prove a specific goal. Conflicts are inevitable. The descriptions of science can not satisfy the desires of the religious. Since that is true, the religious fear the scientists. But it is also true that many scientists fear other scientists.

To quote Feynman again, "[The observations of a scientist] cannot be rough. You have to be very careful... You have to check the observations very carefully, and then recheck them, to be sure that you understand what all the conditions are and that you did not misinterpret what you did." Part of the checking is to compare your observations with all of the well-established laws of chemistry and physics that could possibly apply. Can science explain the disappearance and reappearance of a physical body? Can science predict the effects of such an event? When a great, internally-consistent superstructure of conclusions is based upon scientific-sounding postulates that can not be supported, a scientist can become paranoid about his claims. "Scientific" studies on the Shroud must be unique in the history of science for the high content of easily refutable "theories."

Science demands that we make careful observations and we check them until we can have some confidence in them. Studies on the Shroud of Turin are unique in another way: **It is terribly difficult to check results.** It is an object of devotion, and people love it. It would be unthinkable to do destructive tests on it. Although its custodians have allowed some unconscionable and incompetent sampling and operations to be performed, they are terribly reluctant to allow any careful, rigorous, international scientific studies. One exception was the study made in 1978, and I will describe the observations and critical scientific laws in as much detail as possible. I can not *prove* any hypothesis or even advance any to the status of a scientific theory, but I hope that the reader can draw his or her own conclusions about whether the observations constitute good science and can use them to test his or her own hypotheses.

The primary goal of STURP was to test the hypothesis that the Shroud's image was painted. Secondary goals were to observe the Shroud's technology and composition so that statements could be made on its "authenticity" (whatever that is taken to mean) and possible age. In order to accomplish the goals, many observations were made before, during, and after the Shroud was made available for study in Turin. Facts were also assembled from literature surveys. Some of these facts will be mentioned to illustrate the process of Scientific Method. The observations and data can and should be used to test all hypotheses on the Shroud. No facts can legitimately be ignored.

CHAPTER II: THE FIRE OF 1532

Fortunately for chemists, the Shroud was severely burned during the fire of 1532. The fire provided an excellent "chemical test" on all of the materials of the Shroud. Before it was possible to make direct observations on the cloth, we were able to observe the effects of the fire on the image, apparent blood stains, and clear parts of the cloth by looking at high-resolution photographs.

At the time of the fire, the Shroud was being stored folded in a silver and wood reliquary. That historical information is important, because it enables us to predict the chemical effects of the fire on the cloth. As the fire smoldered into the cloth, oxygen was depleted inside the reliquary. The chemical products of such a pyrolysis are well known. The reactions between those products and every chemical compound you might hypothesize as being involved with what we see on the Shroud can be predicted. We were able to get valuable information before we looked at the cloth in Turin.

If the image had been painted, some colored material had to be added to the cloth. The pigments and vehicles would change in response to heating, the pyrolysis products, and the water used to put the fire out.

The oldest known paintings appeared in prehistoric times (18,000-15,000 BC), and they are found in the caves of France and Spain. They were done in natural materials, e.g., red and yellow ochre and charcoal. There is evidence that the pigments were mixed with animal fat for application to the irregular cave surfaces. Tempera painting appeared early in history. It involves powdered pigments mixed with egg, plant gums, and/or glues. Aside from fresco, tempera was the principal painting medium before the introduction of oil paints.

The Flemish brothers Hubert and Jan van Eyck are generally (probably incorrectly) credited with the invention of oil painting. Their careers are well documented between about 1422 and 1441. They normally worked on canvas that was made from either linen or a linen-cotton blend. It would be extremely unlikely that oil paints had been used to hoax the image during or before the 14th Century; however, we planned observations that would detect such materials. Oils were the favorite vehicles for pigments during the time of the 1532 fire, and they could have been used in an attempt to reproduce the image.

We had to consider the possibility that the famous, money-making relic had totally been destroyed in the fire. It was possible that the original Shroud had cleverly been reproduced, complete with carefully placed scorch marks. After all, the Shroud was not seen publicly between 1532 and 1534.

Some of the greatest artistic geniuses of all time were alive and active during these years. Foremost among them, Michelangelo Buonarotti (1475-1564) had been involved with the defense of the Florentine Republic in 1530, and his whereabouts are not completely documented while the Shroud was out of sight. We considered that he could have been commissioned to reproduce or repair the Shroud after the fire, and his innovative brilliance would have provided a real challenge to scientific observations.

No matter what materials were used to paint the image and/or blood, the temperatures, temperature gradients, pyrolysis products, and water used to extinguish the fire would have changed the chemical composition of most foreign materials.

I even considered the possibility that the image had been painted with a spirit lamp and blowpipe. The chemical composition of such a color would be very difficult to differentiate from natural aging. Hot irons produce slightly different chemical products than does a high-temperature flame (e.g., a blowpipe or welding torch).

Before going to Turin in 1978, we did many experiments on the stability of the painting materials. All of the pigments and vehicles that we could identify in historical documents were tested by applying them to linen and subjecting the samples to different kinds of heating.

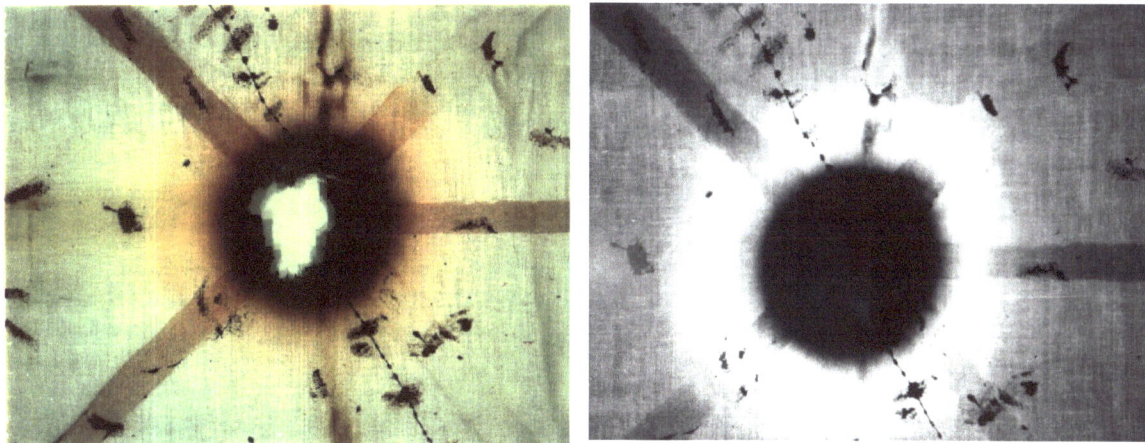

Figure II-1: (Left) Results of one "burn test." Linen streaked with blood and different painting media with and without pigments was heated intensely in the center while it was confined between two stainless-steel plates (1977)

Figure II-2: (Right) The same sample under UV illumination. Condensed cellulose pyrolysis products form an intensely fluorescent ring around the center of heating. Few fluorescent products are produced by pyrolysis in open air.

Based on color photographs, we had expected hematite (Fe_2O_3) to be the most probable pigment that could have been used to paint the image; therefore, the tests we designed emphasized that hypothesis. After the 1978 studies in Turin, Walter McCrone claimed that his microscopic observations proved that the image was a hoax, made by application of hematite to the cloth. He did not try any independent tests to confirm his claim, and it can be dismissed on the basis of several pieces of evidence. However, before McCrone's claims, the heating tests proved that much of the red hematite was reduced to black magnetite (Fe_3O_4) by the pyrolysis products of linen (cellulose, hemicellulose and lignin). We looked for magnetite at scorch-image intersections on the Shroud in 1978, but we could not find any evidence for it.

All paints were changed by heat and/or the chemically reducing and reactive pyrolysis products. Some medieval painting materials become water soluble, and they would have moved with the water that diffused through parts of the cloth as the fire was being extinguished in 1532. Observations of the Shroud in 1978 showed that nothing in the image moved with the water.

When linen is heated, water immediately begins to be desorbed and the linen dries out. As the temperature increases, the cellulose melts with decomposition. Quickly heated and cooled linen shows little black balls where it melted. As it melts, the carbohydrates (cellulose and sugar-based hemicellulose impurities) start to dehydrate chemically. The colored products of dehydration are extremely complex, but they have some well-known chemical properties and structural units.

The color you see in scorched linen or caramelized sugar is a result of interactions of light with what are called conjugated carbon-carbon double bonds. The schematic chemical structure is shown as something like the following: -C=C-C=C-C=C-. These are called "unsaturated" structures. Notice that alternating pairs of carbon atoms have double bonds between them. The more alternating double bonds, the more the color shifts from blue toward red.

Benzene is a symmetrical ring composed of six carbon atoms with three double bonds. Some complex structures containing several benzene rings absorb ultraviolet light and re-emit visible light: they are said to "fluoresce." Some pyrolysis products of linen fluoresce, but the amount that forms depends on the amount of oxygen available during the pyrolysis event. The burning cloth in the sealed reliquary produced a significant amount of fluorescent products around the scorched areas. The fluorescent areas are clearly visible in photographs that were taken with "black light," i.e., ultraviolet illumination. *The image does not fluoresce at all.* If the image involved intense heating, it probably formed in open air.

More detailed chemical studies have shown that the major secondary products of the thermal decomposition of cellulose and other carbohydrates are formaldehyde, carbon monoxide, furfural (2-furaldehyde), hydroxymethylfurfural (5-hydroxymethyl-2-furaldehyde), levulinic acid (4-oxopentanoic acid), and 3-pentenoic-γ-anhydride. Formaldehyde, carbon monoxide, furfural, and hydroxymethylfurfural are potent reducing

12

agents, especially at higher temperatures. They can reduce iron and copper ores, and the salts formed by the reactions of many of the elements used for pigments with organic acids are soluble in water. It was important to observe and analyze the composition of products that had been carried by the water used to extinguish the fire in 1532.

During work on prehistoric pottery firing, I found that cellulose produced relatively more levulinic acid than furfural when compared with most of the plant gums that are used in painting. During the fire, there should have been plenty of organic acids present to react with natural, inorganic pigments.

Observations we made before the trip to Turin in 1978 made us believe that the painting hypothesis was unlikely. This made the project appear much more interesting to a chemist, and we extended our plans to test every hypothesis we could think of at the time.

CHAPTER III: MICROSCOPY

The Shroud was observed through a microscope for many hours in 1978. We had wondered about microbiological attack on the cloth, so that was the first thing we looked for. No evidence for fungal or bacterial attack could be seen.

A large number of photomicrographs were taken by Evans and Pellicori. In the process of making these observations and photomicrographs, they observed the strange, unexpected distribution of the image color on the cloth (figure III-1). The image is the result of many superficial yellow linen fibers. Even Walter McCrone, who only saw fibers I took from the surface with adhesive tape, reported that "Microscopically, the image consists of yellow fibers… and the yellow fibers are the major colored substance in the body image."

During the 1978 observations in Turin, I used a dissecting needle to push some of the individual, superficial, yellow, 10-15-μm-diameter image fibers aside and look under them with the microscope. I could not see colored fibers more than two or perhaps three deep below the top surface of a yarn at the highest part of the weave.

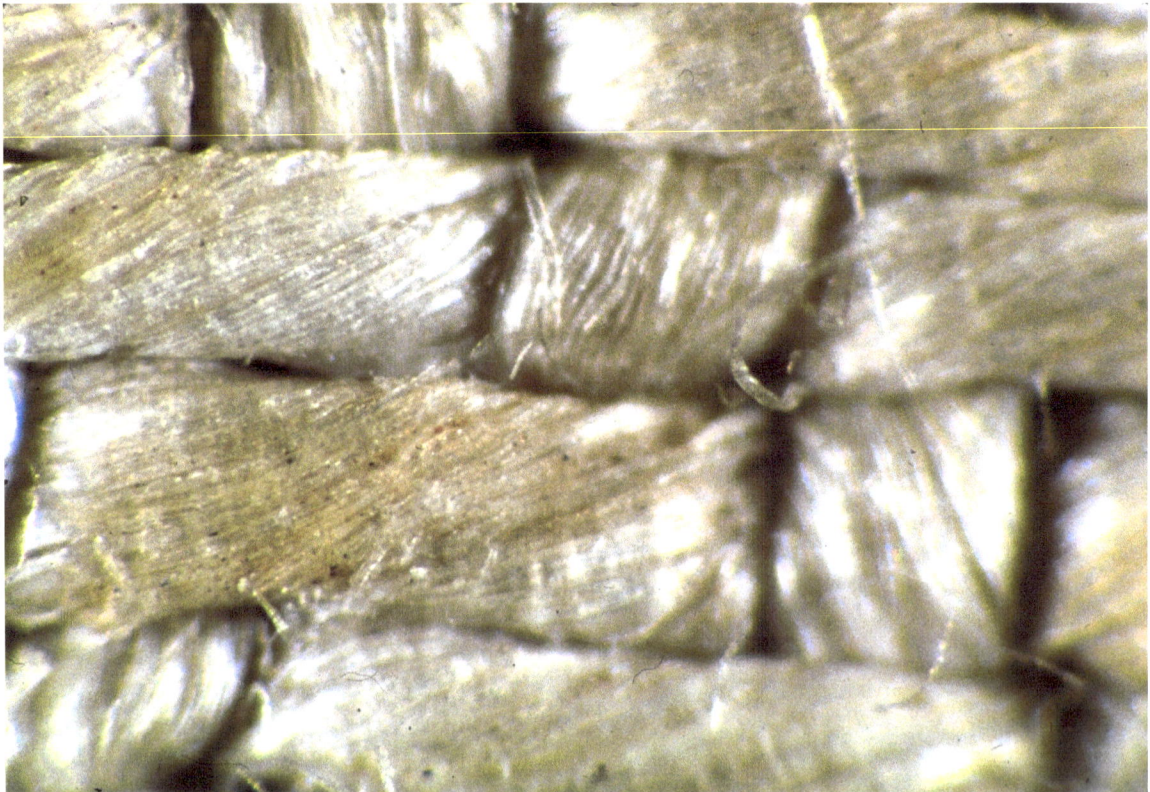

Figure III-1: Photomicrograph of a dark part of the image at the bridge of the nose. The color appears only on the topmost part of the weave, and it does not penetrate very far into the cloth. © 1978 Mark Evans

Since we were making direct observations on the surface of the Shroud, we had to use light that was directed onto the top surface of the cloth and reflected into the microscope. Usual microscope technique uses transmitted light. The incandescent light we used was slightly yellow, and it was extremely difficult to see the light-yellow image fibers. There appeared to be slight differences in the density of color in different image areas, but all of the fibers I could identify as image fibers seemed to be very similar. The statement that all of the image fibers are nearly identical may be a result of viewing conditions as much as similarities among fibers. It is possible that I could see only fibers of a specific color density.

The color density seen in any area of the image appears primarily to be a function of the number of colored fibers per unit area rather than a significant difference in the density of the color of the fibers. This observation was puzzling, and we called it the "half-tone" effect. After observing hundreds to thousands of fibers on sampling tapes that were taken in 1978, I still believe that there is a very limited range of image-fiber color densities. The half-tone effect is probably mostly true.

No fibers in a pure image area were cemented together by any foreign material, and there were no liquid meniscus marks. The material that resembled blood stains was quite different. These facts seemed to eliminate any image-formation hypothesis that was based solely on the flow of a liquid into the cloth. This also suggests that, if a body was involved in image formation, it was dry at the time the color formed.

I will not discuss details on the chemistry of the blood stains (figure III-2). We tested them to make certain that they contained real blood and not a red pigment. They definitely contain the oxygen-transporting porphyrin of real blood, heme. Professor Alan Adler invested considerable effort on a study of the blood, and other members of STURP did many different microchemical tests for proteins. The only protein we found was in visible blood areas; however, all blood areas tested gave protein tests. There were no "blood images" that did not show the protein test. We reject the idea of non-contact blood "images" that did not involve the physical transfer of blood to the cloth.

There is no evidence for significant amounts of any of the many pigments and/or dyes that could have been used to paint or touch up the blood stains. We had considered and studied Tyrian purple (6, 6'-dibromoindigo) and Madder root dye on an aluminum and/or chromium mordant as well as cinnabar (mercuric sulfide) and ferric oxide pigments.

Human blood has always been easy to obtain, if you did not seriously consider "human values." Animal blood has always been cheap. The presence of blood on the Shroud proves nothing. However, the presence of some foreign material other than blood that was used to give the appearance of blood would be conclusive evidence of hoaxing or tampering.

Direct microscopy on the surface of the Shroud showed that the red stains were composed of a foreign material that stuck to the surface of the linen fibers. It was also easy to observe areas near red blood spots where the fibers were coated with a golden, transparent material. We assumed that it was blood serum, and later chemical tests on recovered samples proved that those areas were rich in proteins. The material was almost certainly blood serum.

It was interesting to notice that, where serum had flowed through the cloth in an image area, there seemed to be no image-colored fibers below it. It appeared that the blood had protected the cloth from whatever mechanism had produced the image.

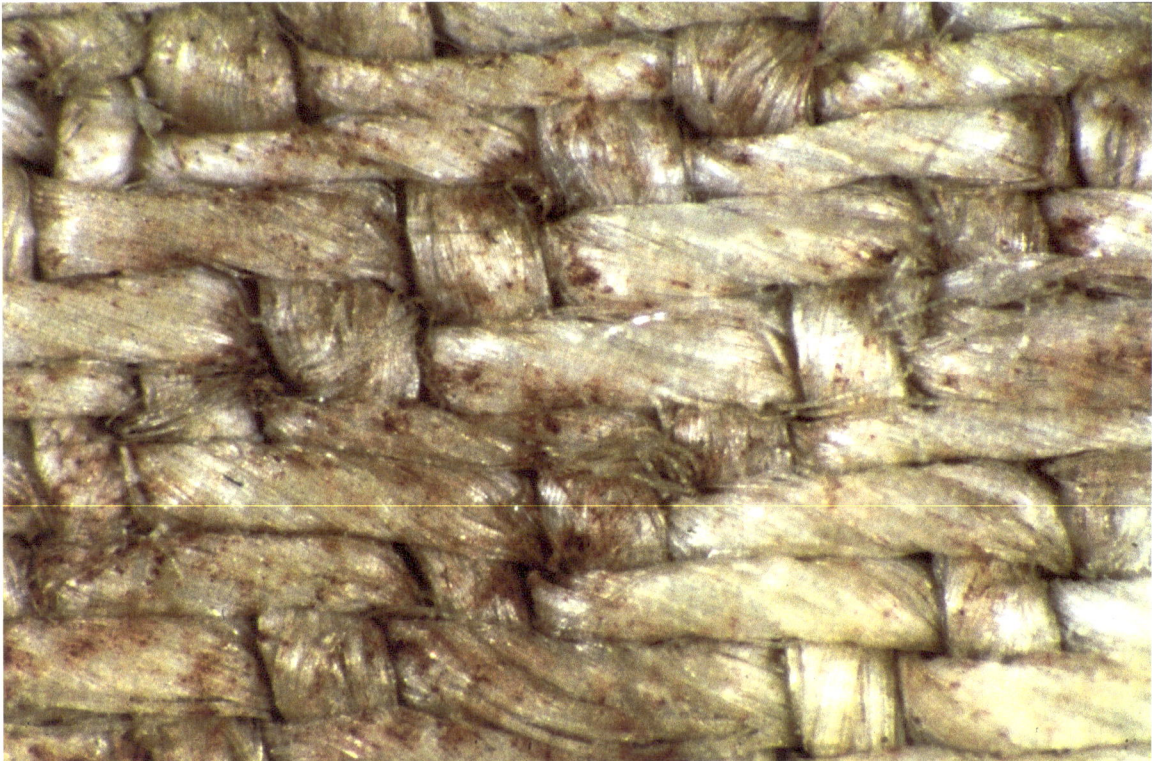

Figure III-2: Photomicrograph of a blood area. Red flecks can be seen adhering to the linen fibers and the material has soaked into the cloth. © 1978 Mark Evans

Diffusion of gaseous reactants or dyes into the cloth would have produced a color gradient (darker on the surface, lighter at depth). The same can be said for a surface scorch, the color would penetrate farther into the cloth in darker areas. This was not the case with the Shroud image.

CHAPTER IV: PHOTOGRAPHY

The first photographs of the Shroud were taken in 1898 by Secondo Pia. He used a special camera and large glass plates that were coated with a photosensitive emulsion. Inadvertently, he performed a scientific experiment. His plates were much more sensitive to blue light than are our eyes or modern orthochromatic film. Comparisons between his photographs and later orthochromatic ones gave some preliminary indications of the kind of information we could get by making spectrometric measurements. They were a great help in planning the chemical observations

Pia's photographs showed the Shroud in an entirely "new light." Details of the very faint image could be seen that were invisible to the unaided eye. We would say that his photographs were "contrast enhanced (figure IV-1)."

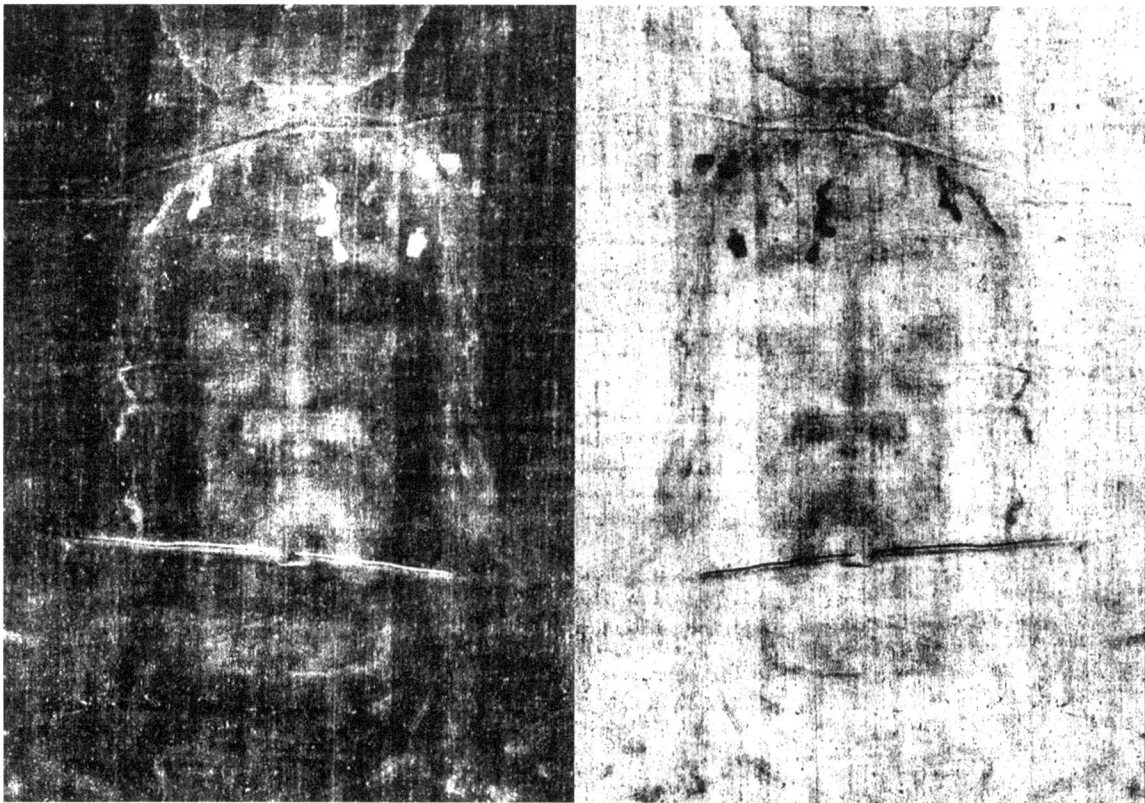

Figure IV-1: Contrast-enhanced photographs of the facial image. Left, negative; right, positive. © 1978 Barrie M. Schwortz

When he first looked at his photographic plates, Pia was startled by what he observed. Nobody had ever seen the image like that before. The negative actually looked like a photographic positive. This initiated a massive amount of pseudoscience that postulated different, usually miraculous, "photographic" processes to explain image formation.

Aside from the emotional impact of the enhanced image, some important objective observations can be made. I believe that the most important is that bands of different-colored yarn can be observed in the weave of the cloth. A few light vertical streaks can easily be seen on the negative view, and some anomalously light areas can be seen on the positive presentation. Similar bands of light and dark can be seen visually with some difficulty on the Shroud, but they appear much more clearly when contrast is enhanced. They appear in both the warp and the weft. The observations of bands of color agree with historical reports on the methods used to produce ancient linen. They indicate a very mild bleaching technique, unlike that used after the last crusade in AD 1291.

According to Pliny the Elder, who wrote his *Natural History* about AD 77, ancient linen yarn was spun by hand on a spindle whorl. When the spindle was full, the spinner prepared a hank of yarn for bleaching by the fuller. Each hank of yarn was bleached separately, and each was a little different; indeed, different parts of the same hank show slightly different colors, a little like variegated yarn. This effect can be observed on the Shroud.

A conservator at Turin's Museum of Egyptology, Anna Maria Donadoni, pointed out locations where batches of yarn ended in the weft and new yarn had been inserted in order to continue weaving. The yarn ends were laid side by side, and the weave was compressed with the comb. The overlaps are often visible, even in high-resolution x-ray photographs. When an overlap is observed, the color usually changes. The color of the Shroud is not simply a result of changes in pure cellulose (linen). The bands of color prove that there were, and still are, impurities on the surface of the yarn. This helps confirm the ancient nature of the linen-production technology. It also suggests that impurities of interest in the context of image formation existed on the cloth's surface after it was produced.

The warp yarn was protected with starch during the weaving process, making the cloth stiff. The final cloth was washed with "struthium," *Saponaria officinalis*, to make it more supple. There is no question but that traces of starch components would be left on the cloth after washing.

Saponaria hydrolyzes to produce some aglycones that are fluorescent, and the non-image part of the Shroud is weakly fluorescent. The use of *S. officinalis* to wash the cloth could explain the fluorescence of the background. The image either filters or quenches that fluorescence.

Medieval linen was bleached as the whole cloth. Most commercial bleaching took place in "bleach fields" in the Low Countries, the genesis of the name "Holland cloth" for the medieval backing on the Shroud. Considerable material was lost during the bleaching process, and the newer linens are less dense than ancient linens, as could be seen by comparing the Holland cloth and patches with the main part of the Shroud. The newer linens are also homogeneous. They do not show bands of different-colored yarn in the weave as the Shroud does.

18

Incidentally, we found that *Saponaria officinalis* solutions are hemolytic: they break the membranes of red blood cells and release the red hemoglobin. Hemolysis is used to determine the hemoglobin content of blood. Whole blood darkens as it ages on cloth; however, the blood spots on the Shroud are still quite red after centuries of known history. Diane Soran (deceased) of Los Alamos tested hemolysis on *Saponaria*-washed cloth before we went to Turin. The blood is still red on those 25-year-old samples although the blood on non-*Saponaria*-washed control samples is black. This fact might help confirm that ancient technology was used to produce the cloth.

Vern Miller and Sam Pellicori of STURP photographed the Shroud in absolute, darkroom-type darkness as it was being illuminated with a pure "black-light," ultraviolet source. The source was completely filtered to eliminate any visible light. The camera was carefully filtered to remove all traces of ultraviolet light that could affect the film. The only light that can be seen on their UV photographs has been produced by fluorescence, the visible light produced by specific interactions between the energetic ultraviolet light rays and the chemical components on the Shroud.

Figure IV-2: Ultraviolet-fluorescence photograph of the image hands. A few highly-fluorescent, modern fibers can be seen on the surface. © 1978 Vernon Miller

You can easily see the bands of different yarns in the cloth in the UV photographs (figure IV-2). The non-image part of the cloth fluoresces faintly, giving a soft glow to the

19

cloth. Some bands of yarn are lighter and some darker, making the appearance somewhat like a plaid. The fact that different yarns show different amounts of fluorescence is important. Differences in fluorescence prove differences in chemical composition. The yarn used to weave the Shroud was not one homogeneous batch, and the entire cloth was not bleached at the same time by the same process.

The image appears dark on the fluorescent background: The image does not fluoresce, but it appears to quench or filter the fluorescence of the background cloth. Blood spots appear almost black in fluorescence photographs.

Early observations of the Shroud that we read before the studies in Turin reported that the blood fluoresced. That would have been remarkable and could have had important historical implications. Persons with porphyria have high concentrations of porphyrins in their blood and urine, and their blood fluoresces. The UV photographs and careful spectrometric observations eliminated this possibility.

Sam Pellicori reported that the different patches on the Shroud fluoresced differently, but we were never able to pursue that observation. He also reported that the margins of the scorches fluoresced in the green, entirely different than the background of the Shroud.

When blood spots appear on an image area, lighter "halos" can be seen around most of them. These areas were the same as had been observed microscopically to be coated with presumed blood serum. It appeared that blood serum had prevented image formation and had preserved the background fluorescence of the original cloth.

Where darker bands of yarn intersect image areas, the image is darker. Where lighter bands intersect an image area, the image appears lighter. This proves that the image color is not solely a result of reactions in the cellulose of the linen. Something on the surface of the different batches of yarn produced color and/or accelerated color formation. This observation is extremely important when tests are being made on image-formation hypotheses. If image color is not simply a result of the coloring of the cellulose of the linen fibers, image formation must be a much more complex process than we originally thought.

Mottern, London, and Morris of STURP took duplicate, low-energy, high-resolution x-ray photographs of the entire Shroud. The resulting photographs showed differences among the densities of the background cloth, blood stains, scorches, and water stains. However, *the image was totally invisible on the x ray photographs.* No material that had a significant electron density had been added to the cloth in order to produce the image color.

CHAPTER V: SAMPLES

Adhesive-tape sampling:

I took samples from the surface of the Shroud in 1978, using tape made specifically for the task by Ronald Youngquist of Minnesota Mining and Manufacturing, Inc. The

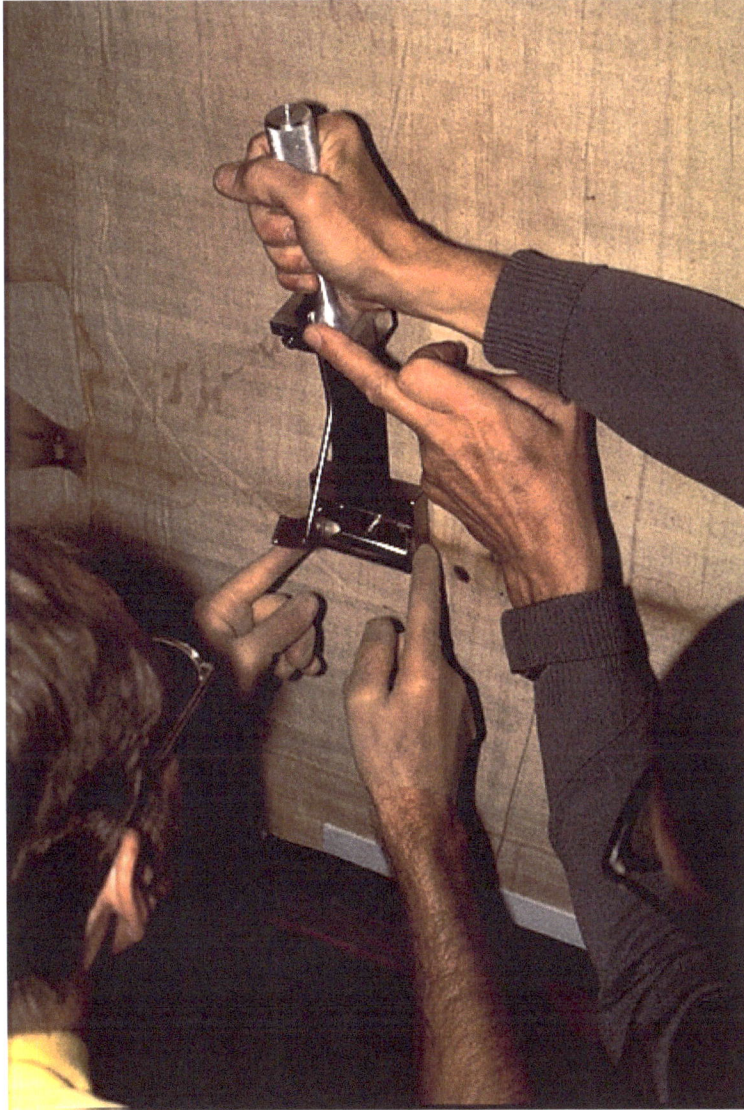

Figure V-1: Applying a tape segment to the Shroud with a force-calibrated roller. © 1978 Barrie M. Schwortz

backing and adhesive of the tape were chosen to be amorphous; i.e., they did not give any birefringence colors under crossed polarizers when viewed under a petrographic microscope. The adhesive was a pure hydrocarbon that did not contain any liquid fractions. It could not contaminate the Shroud, and the inert adhesive enabled many types of chemical tests to be made directly on the tape's surface.

Our efforts and the expense of obtaining an inert sampling tape were "over-kill," because Max Frei was allowed to sample for pollens with commercial tape. It is interesting that Professor Alan Adler observed dark marks on the Shroud at all of Frei's 1978 sample sites when he was given a rare opportunity to examine the cloth a year before his death. These were most likely the result of dirt adhering to remnants of the adhesive left on the cloth by Frei's commercial tape.

The tapes were applied to the surface of the Shroud with a force-calibrated applicator (Fig. V-1). My previous experience with tape sampling for chemical analyses indicated that controlling the force enabled semi-quantitative comparisons among samples. The use of a reproducible force also allowed estimation of the comparative physical states of sampling areas. Different painting media adhere to linen with different tenacity, and chemically degraded linen has less physical strength. We hoped to observe such features without damaging the Shroud.

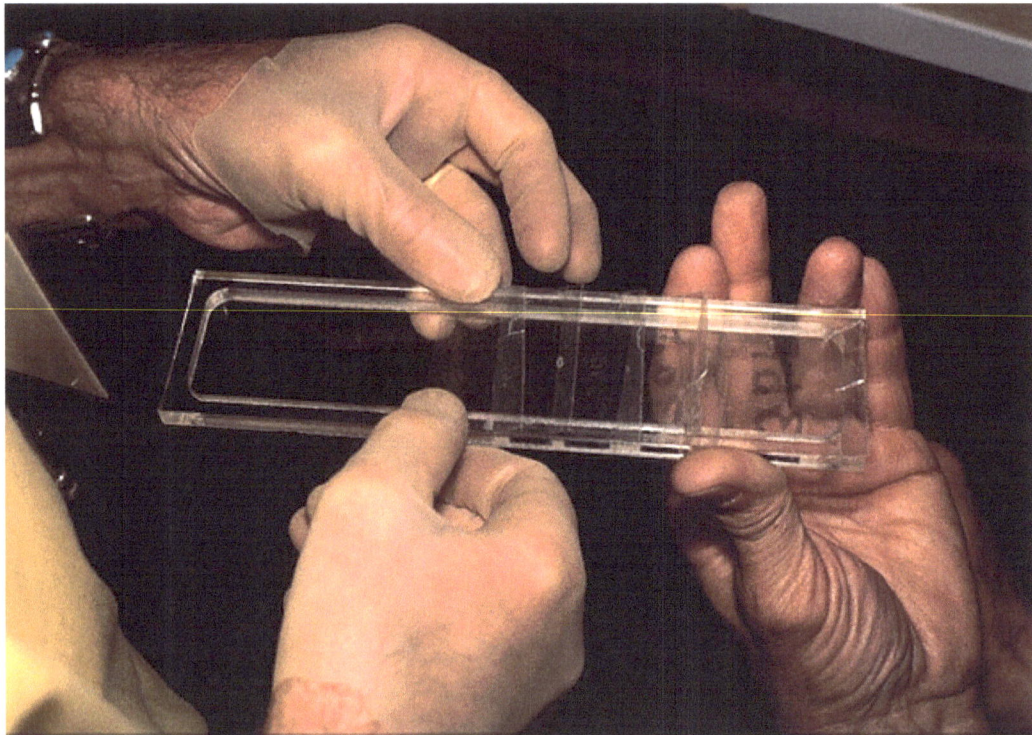

Figure V-2: Sampling tapes with charged fibers pointing straight up from the adhesive being placed in a protective frame. © 1978 Barrie M. Schwortz

As anyone who has pulled tape off its roll or other surfaces in the dark has observed, a large static charge is created. You can see flashes of sparks, and the flashes are intense enough to expose photographic film. This static charge was applied to the problem of keeping fibers pristine (Fig. V-2). The linen fibers are good insulators, and they will accept and retain a static charge. This charge has exactly the same effect on the linen fibers that is seen in an electroscope: the fibers were repelled from the adhesive surface. Since they did not touch any other surface, I had planned to remove them individually from the surface with clean forceps and be able to do all of the different chemical analyses we had planned without any interference. This protection from contamination was basic to the pyrolysis/mass-spectral analyses we had planned (Chapter VIII).

In order to protect the charged fibers from dust, I designed a gasketed, sealed box that could be flushed with an inert atmosphere. The inert atmosphere was an attempt to avoid any unwanted reactions at stressed locations on fibers. Walter McCrone claimed that the box contaminated the tapes with jewelers' rouge that had been used to polish the plastic. The plastic had been cleaned thoroughly and sealed in a clean-room after it was assembled. It was not opened again until the fresh samples were placed in it in Turin.

Walter McCrone had taken part in the 1977 meeting at which STURP was organized. He badly wanted to develop a micro-radiocarbon system for dating the Shroud. I had known Walter since the 1950s, and I considered him to be both an ethical scientist and one of the world's best microscopists. We agreed to share the work on the tape samples.

Walter later mightily irritated the custodians of the Shroud by going over their heads to talk to King Umberto, who was the surviving member of the Savoy family and who owned the Shroud. He was consequently refused admission to the experimental area in Turin, and he never became a formal STURP team member. He had nothing to do with the sampling procedure or equipment.

Walter and I had made an agreement to share uncontaminated samples before any chemistry was done on them. The first step in making observations would be direct, low-power microscopy on the original tapes. I opened the sample box in a clean room on 27 October 1978. The adhesive sides of the tapes were protected in their frame, but I could look through the tape with a low-power microscope. Sample 3EF from the wrist blood spot showed a number of red spots. It was marked to reserve for blood testing. Sample 3AF from the middle finger showed a large number of yellow fibers. It was marked for image testing. Many other image-area samples showed yellowish fibers. The tape samples appeared to be perfect for our purposes, and we celebrated.

I had kept a few tape samples open in the experimental room in Turin to check for unexpected particulates that might be drifting around. Those tapes appeared to be nearly free of fibers.

The next step would be to observe small segments of the tapes at higher magnification. I did not want to open the sealed box more than necessary. I let Walter have the box of samples and take it to his laboratory. He agreed to open the box in a clean-room, and he promised that he would do nothing but cut small sections of the tapes for microscopy. He said: "You know how little that will take." He also promised to maintain the "chain of evidence" for the samples, allowing no unauthorized access to them. He paid no attention to his promises: He nearly destroyed the value of the carefully prepared samples.

Walter McCrone took it upon himself to stick all of the samples down to microscope slides. He did not reserve *any* samples in a pristine state. Even worse than that, he immediately found that he could not tolerate the optical effects of a thick slide. That is amazing performance for a "great" microscopist. Consequently, he pulled the tapes off of the slides and stuck all of them down to microscope cover slips. This destroyed much of the physical evidence we had sought. Some fibers can now be seen to have been broken during these transfers, and thin coatings were often pulled off of fibers' surfaces.. This exacerbated contamination with adhesive, and it also initiated crystallization of the adhesive and amorphous tape at a higher rate than would have been necessary. The original tape is still amorphous, but the samples mishandled by McCrone are crystallizing. McCrone's failure to follow protocols and his abuse of the samples were unconscionable.

Fortunately, we had chosen the adhesive to provide for quantitative removal from the samples. It required much meticulous work to get around the damage caused by McCrone, and much information was simply destroyed by his actions.

Another final disaster befell the tapes. I had transferred them to Professor Alan Adler for additional microchemistry and identification of the blood. When Al died unexpectedly on 12 June 2000, his wife sent the samples to Turin. There is still much scientific information available in those tape samples; however, I have no idea where they are or how they are being kept and preserved. The authorities in Turin will not answer any inquiries.

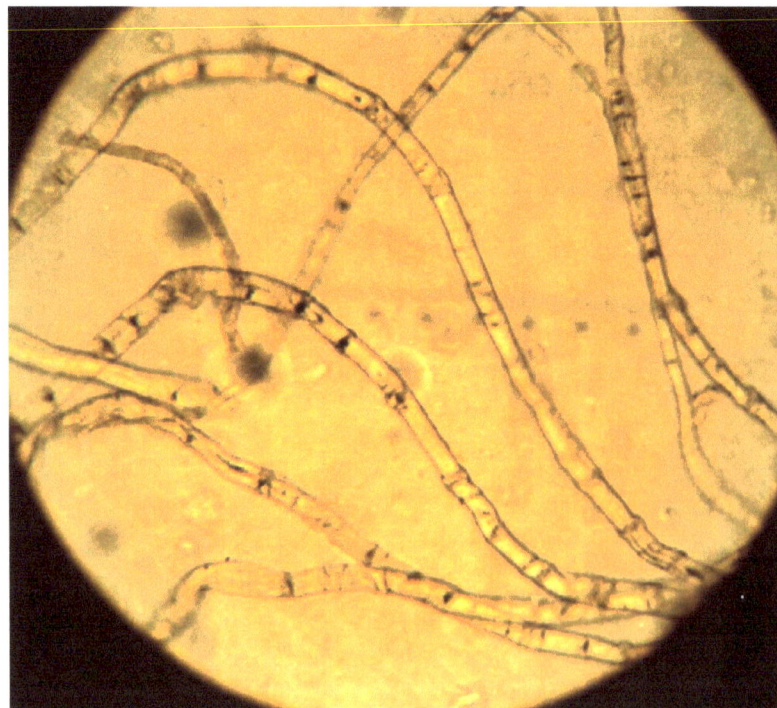

Figure V-3: Yellow image fibers (400X), showing black lignin at growth nodes. The tape background has yellowed with age, reducing color contrast.

Everything connected with the Shroud is desperately desired by some persons. They consider everything that has touched it to be a relic. Representatives of the Cardinal actually jostled each other to get rubber gloves, cotton gloves, and thumb tacks out of the waste basket in the experimental room in Turin. I have to assume that the few poor remaining "scientific" tape samples are framed and hanging on someone's wall.

To maximize safety to the Shroud, I started the sampling procedure by pulling a tape from a patch rather than the main part of the Shroud. This frightened me, because it pulled so hard that I was afraid the cloth would tear. I reduced the application force to a level where tapes pulled easily from the patches. When I took a tape from a non-image area of the Shroud, I found that it pulled much more easily than tapes pulled from the patches. The large difference in ease of pulling tapes from the surface made me decide to use the applicator to measure the force required to remove tapes. Tapes pulled from darker body-image areas with extreme ease: I could barely measure the pulling force.

Image fibers on the sampling tapes are normally yellow. A few are a darker yellow-brown. Many fibers from blood areas are a deep red.

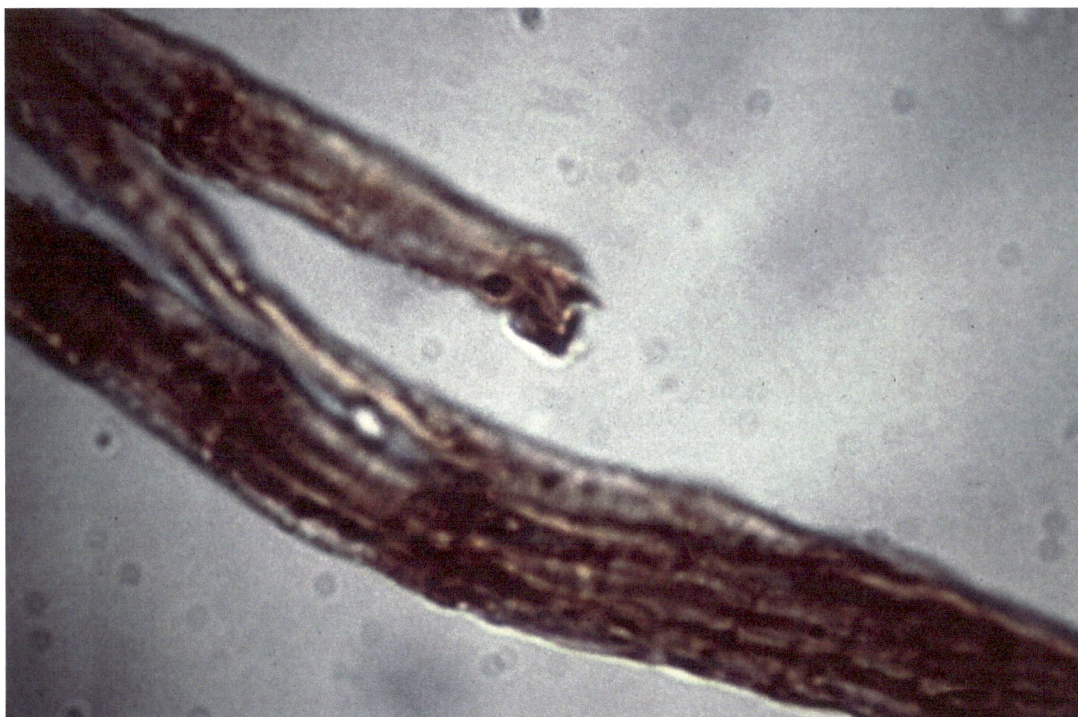

Figure V-4: A fiber from sample tape 6AF, taken from the "lance wound" on the front image. It gave an intense positive test for proteins.

Raes and Radiocarbon Samples:
Professor Gilbert Raes of the Ghent Institute of Textile Technology cut a small sample from the cloth in 1973. He found that the samples contained cotton, and he reported that the cotton was an ancient Near Eastern variety, *Gossypium herbaceum,* on the basis of

the distance between reversals in the tape-shaped fibers (about eight per centimeter). I can not confirm the identification of the cotton variety; however, I can confirm the presence of cotton in the Raes sample. The cotton is important.

Cotton was almost unknown in Europe until about AD 1350, when "there was widespread belief that it was the fleece of miniature sheep that lived in trees." Crusaders helped spread knowledge of cotton through Europe. There were still legal disputes over whether cotton was a kind of linen as late as AD 1631.

In 1980, I received 14 segments of yarn from the 1973 textile sample from Professor Luigi Gonella, Department of Physics, Turin Polytechnic. I now have them numbered and identified as the "Raes threads." I also received an envelope that was supposed to contain one of the Zina threads that had been cut out of a blood area. There was very little sample in the envelope: The Zina sample had been picked to pieces. Some red flecks could be seen on the outside of the inner wrapping, but they could not legitimately be associated with a Shroud sample. Such evidence would never hold up in court.

The results of analyses on the Raes sample made it appear that the radiocarbon sample that was cut for dating in 1988 was not valid (Chapter IX). This naturally caused another controversy. The only way to settle the validity question was to obtain some threads from the authentic radiocarbon sample. However, in dealing with Shroud problems, every sample must be documented without any gaps. I finally received a documented sample of radiocarbon yarn segments, both warp and weft, on 12 December 2003. Professor Luigi Gonella had pulled these segments from "the center of the sample" in 1988 at the time of sampling. Their history was known from that date.

There has been a strong tendency for people to steal samples as relics. This has provided a serious impediment to scientific observations. There is also a strong tendency to pick samples to pieces. The Raes sample was cut as woven cloth, but the threads I got had been picked apart. I could not be sure which yarn segments came from the warp and which might have been weft. It was impossible to observe cloth structure or possible interactions between yarn segments. I have called this the PPSP factor, people-picking-samples-to-pieces. The Shroud project has been the most frustrating "scientific" study in history.

One redeeming feature of the project has been the telephone calls and letters. One man with a deep Southern accent called to tell me how the image formed. He said: "Hey Bud. You know they have them titty flies in North Afriker. If they smell a dead body under a piece of cloth, they will poke their little noses through the cloth to get a bite. Now, you jest look at that cloth and see if there ain't a bunch of fly specks making that image." I was out of action rolling on the floor and I couldn't answer him.

One man wrote me a lot of letters from a federal prison. When he didn't like my lack of responses, he promised to "come see me" when he got out. Shroud studies can make life exciting.

Another man sent me many letters on image hypotheses, most written on envelopes. He used a name that was current in news papers at the time as a sought serial killer. I turned the envelopes over to the FBI and never heard from him again. He didn't really have much insight into science - - - just death.

CHAPTER VI: DISTRIBUTION OF IMAGE COLOR

Many observers have remarked on the surprising distribution of image color on the Shroud. The colored fibers appear only on the very topmost parts of the weave, and significant parts of the weave in an image area are not colored at all between the spots of color. The color is distributed in a *discontinuous* pattern in all image areas. The effect can be seen clearly in the photomicrographs that were taken by Evans and Pellicori (Chapter III).

In addition to the discontinuous distribution of the color, the color does not penetrate the cloth to a significant depth in most places. When I looked at image areas through a microscope while pushing surface fibers apart with a needle, I could not be sure that colored fibers existed more than two deep in the dense image area at the tip of the nose. In some other image areas, it was nearly impossible to see a second level of colored fibers.

When we looked at the back of the cloth in 1978, we could not see any image; however, the photographs of the back surface that were taken in June and July of 2002 show that light image color appears in the area of the hair. This is a very important observation for discussions on image formation hypotheses. Whatever mechanism produced the image color did not significantly penetrate the cloth above areas of skin but did above the hair.

I consider the slight penetration of the color in all areas but the hair to be one of the most important observations with regard to developing image-formation hypotheses. It is also one of the most obscure to explain.

Liquids and gases penetrate cloth. Liquids migrate through fibrous materials by capillary flow. The thickness of the liquid film depends on the surface tension of the liquid compared with the surface energy of the solid. When the liquid films on adjoining fibers touch, the liquid layers coalesce like rain drops on a window pane. The "hour-glass-shaped" liquid bridge between fibers has the same properties as the meniscus you see at the water-glass interface in a drinking glass. When the liquid dries, the concentration of dissolved materials increases as the volume of liquid decreases. The last liquid to evaporate leaves a heavy load of the most soluble materials. The pattern left by materials dissolved in a liquid is characteristic. We could not observe any such patterns in image areas on the Shroud, and we concluded that the image-formation process did not involve capillary flow of a colored or reactive liquid.

Reactive gases diffuse through cloth (or any fibrous material), they react with the surface as they diffuse, and they adsorb to the surface in predictable amounts. The Langmuir Adsorption Isotherm allows you to predict amounts adsorbed. Because they react with the solid, they are depleted as they diffuse through the cloth. When they produce a color, the most dense color is at the surface, but it does not stop at the surface. You see a color gradient as you look into the interior of a fibrous sample. The color at the surface of the Shroud appears to stop abruptly at or near the surface. When the Method of Multiple Working Hypotheses is used as part of Scientific Method, you try to brain-storm as many ideas for how something happened as possible. If we rule out image formation as a result of liquid flow only and as a result of gas diffusion only, we may find that we are left with a combination of the two.

Heat and radiation of sufficient intensity to color cellulose all of the way through the thickness of the cloth would not be limited to producing a color on the back of the cloth in the area of the hair. We would expect to see color in the center of the cloth. We do not. The lower density of the hair makes it unlikely that large amounts of either heat or radiation would have been produced in the hair. This suggests that vapor diffusion was involved in image formation, because any fibrous mat, including hair, reduces the rate of diffusion of gases (see Chapter XII). Fiber mats are used for insulation, because they reduce gas diffusion and heat transfer by convection.

Observations of fibers from many different image areas show that their colors and color densities are very similar, and we could not observe any difference between frontal and dorsal image fibers. The assumed pressure of a body on the cloth seemed to have no effect on image formation. The frontal and dorsal images show the same color density, distribution, and penetration.

When gases diffuse into one another, their rates depend on their densities, which depend on their molecular weights. Their rates of diffusion can be predicted by use of Graham's Law of Diffusion: "The rates (v) of diffusion of two gases into one another are inversely proportional to the square roots of the densities of the gases or their molecular weights."

$$\frac{v_1}{v_2} = \frac{\sqrt{d_2}}{\sqrt{d_1}}$$

A heavy gas diffuses more slowly than a light gas (see Chapter XII). A heavy gas will be retarded by a mat of hair more than a light gas and its concentration will increase relative to the lighter gas.

When a heavier-than-air foreign gas is diffusing into air in a fibrous mat, the concentration of the gas will increase in the hair. More gas will diffuse through the pores of the cloth in the area of the hair. Such a mechanism would explain why the hair is clearly visible in the image and why it is visible on the back of the cloth. The observation of image color at the location of the hair on the back side of the cloth strongly suggests that a gas heavier than air was involved in image formation.

If diffusion is important in image formation, several other phenomena are also required to explain image formation. Again, the image-formation process begins to show complexity. Something that produces heavy, reactive gases must be involved in the process. The visual evidence indicates that the thing that produced reactive gases had the characteristics of a human body complete with hair, a beard, and a moustache. This assumption strongly suggests that the object was a real body and that the body was dead and starting to decompose.

The gases produced by a decomposing body are extremely reactive chemically: many are amines (organic compounds that contain a -NH_2 group). The amines have characteristically nasty odors, and they are responsible for the putrid smell of rotting flesh.

Decomposing bodies start producing ammonia (NH_3) in the lungs quite soon after death, and the ammonia diffuses outward through the nose and mouth. Ammonia is lighter than air, and it diffuses rapidly. The rate of production of ammonia decreases with time after death.

Within a few hours, depending on weather conditions, a body starts to produce heavier amines in its tissues, e.g., putrescine (1,4-diaminobutane), and cadaverine (1,5-diaminopentane). These amines are much heavier than air, and they diffuse relatively slowly. The early appearance and rapid diffusion of low-molecular-weight ammonia from the nose and mouth might help explain the greater amount of image color between the nose and mouth, in the beard, and into the nearby hair. It will also diffuse through the cloth more quickly and reach the back side of the cloth in greater concentration. Ammonia will diffuse about 20 cm through air while cadaverine is diffusing only 6 cm.

The diffusion properties of gases suggest an hypothesis for image formation that involves amines and an impurity on the cloth, and it might help explain the rather good resolution of the body image. Some gases must be produced by the body, and some color-producing reactions must occur on the cloth as a result of interactions between the gases and the cloth.

An interesting corollary of hypothetical diffusion/reaction relationships is the fact that image resolution will be much better when the reactive amines appear slowly. If the amines are reacting with the cloth as they diffuse, their concentrations decrease with both time and distance. If all of the decomposition products were to appear at once, resolution would be poor. Reaction rates could not keep up with diffusion.

Another important principle that can affect image resolution is mixing in the gas phase. Everybody has seen a shimmering mirage. It is caused by differences in air density between hot earth and cool air above the ground. Hot air rises, and this causes turbulence, as any air traveler knows. The same factors apply to the gas between a warm body and a cool cloth that is draped over it. Little circular currents called "convection cells" are established between a warm surface and a cooler one. The rising warm air hits the cooler surface, is cooled, becomes heavier, and sinks toward the warm surface again. Convection cells act to mix gases from either surface with the air. Convection cells are smaller when the distance between the hot and cold surfaces is less. Convection will decrease image resolution. For best resolution, the decomposition products of a body would have to be emitted slowly with the body at the same temperature as the cloth. Little decomposition is required for a trained cadaver-searching dog to find a body, but the probability of finding it from a significant distance increases within a few hours after death as products diffuse farther from the body.

The facts fit together. It takes a few hours for a body to cool after death, and it takes a few hours for heavy amines to be produced in significant amounts. Some image-formation experiments can easily demonstrate how quickly convection cells and vapor concentrations increase with temperature, resulting in decreased image resolution. The vapor pressure of a liquid heavy amine increases rapidly as the temperature goes up. I could not get good resolution when I used a 40°C "body" that was coated with a film of heavy amines.

Image characteristics indicate that ammonia diffused from the nose and mouth, suggesting that the body was wrapped fairly soon after death. The body image has good resolution, suggesting that the heavy amines appeared slowly at lower temperatures.

Many shrouds have been observed in archaeological contexts, and some of them show partial images. The University of Tennessee maintains an experimental area where observations are made on decomposing corpses. They find that flies lay their eggs in wounds on dead bodies, and maggots appear before 30 hours at about 23°C. This approximates the time required for liquid decomposition products to begin to appear on the surface of a body. We could not find any evidence for the migration of liquid decomposition products through the cloth; therefore, the cloth could not have been in contact with the body for very long. These observations also agree with the observations of Dr. Fred Zugibe, a medical investigator, who proposes that the body had been washed on the basis of the blood-flow characteristics.

As mentioned in the chapter on photography, the density of the color of an image area reflects the changes in color density seen in the bands of different color. The image color is not simply a result of changes in the cellulose (linen). Pure cellulose is relatively hard to color by chemical means, but many common impurities on cloth can be colored much more easily. Most of the components of crude starch are carbohydrates (sugars and lower-molecular-weight polysaccharides) that are closely related to cellulose, but they can be quite easy to color by either "caramelization" (heating) or reactions with amines.

Microscopy proves that image fibers and scorch fibers are quite different in structure and composition. The distribution of color is different, even at the level of single fibers. The image was *not* formed by scorching the linen fibers.

When viewed in parallel light under a microscope, a scorched fiber (figure VI-1) is colored through its entire diameter, and the medulla (a tubular void down the middle of the fiber) usually appears to be darker than the mass of the fiber as a result of reactions at its surface and its shorter radius of curvature. The medullas of image fibers do not show any coloration or charring (figure VI-2). Image-fiber medullas are usually clean and colorless.

When cellulose fibers are heated enough to color them, whether by conduction, convection, or radiation of any kind, water is eliminated from the structure (the cellulose is

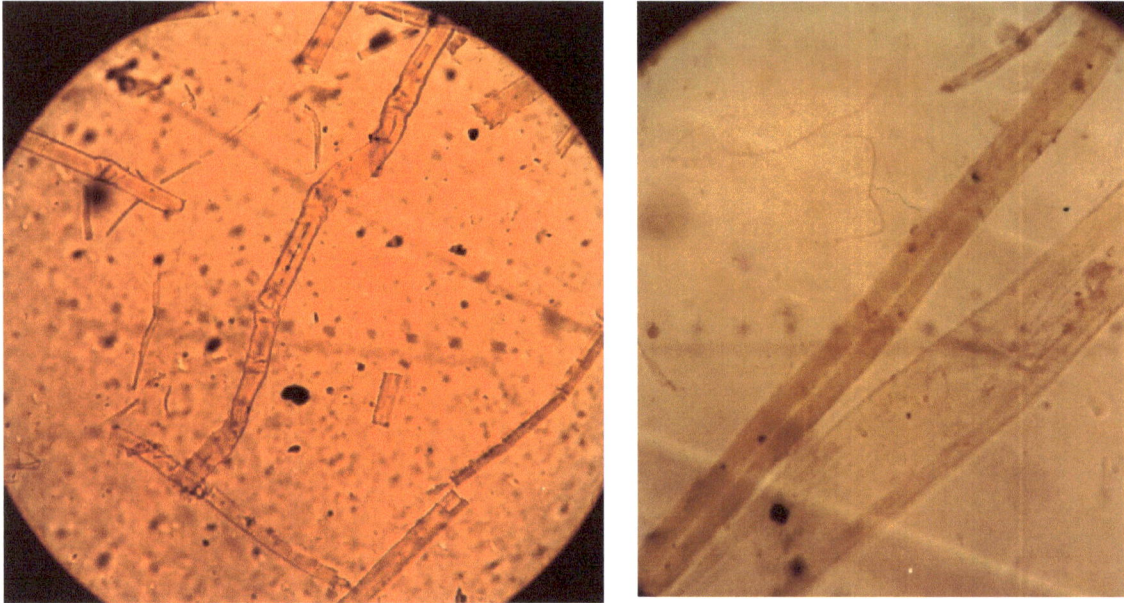

Figure VI-1: (Left) A lightly scorched fiber. The medulla shows darker coloration than the rest of the fiber (X200).

Figure VI-2: (Right) An unusually deeply colored image fiber (X400). The medulla is completely colorless.

"dehydrated"). When water is eliminated, C-OH chemical bonds are broken. The C• free radicals formed are extremely reactive, and they will combine with any material in their vicinity. In cellulose, other parts of the cellulose chains may be the closest reactants. The chains *crosslink*.

Cellulose molecules are folded back and forth in a fairly regular arrangement, and they show the properties of crystallinity. This is called a "fibrillar structure." When you rotate the stage of a petrographic microscope with crossed polarizers while looking at a linen fiber, straight lengths change from black to colored every 45°. The fiber is birefringent and has an ordered structure.

When cellulose starts to scorch (dehydrate and crosslink), its characteristic crystal structure becomes progressively more chaotic. Its birefringence changes and not all parts of a straight fiber go through clear transitions from dark to light at the same angle. Zones of order get smaller and smaller. It finally takes on the appearance of a pseudomorph and just scatters light. A significantly scorched fiber does not change color as the stage is rotated between crossed polarizers.

It is easy to measure the index of refraction of a material and observe changes in its crystallinity with a petrographic microscope. Unfortunately, it is much harder to make measurements on the sampling tapes now that they have started to crystallize as a result of McCrone's damage.

Observations must be made on the specific fibers that reach extinction at the same angle as the tape (while everything is black). The index of refraction of a normal linen fiber parallel to its length is nearly identical to that of the adhesive on the sampling tapes (it nearly disappears). The index across the fiber is appreciably lower than the adhesive. The indexes of refraction and crystallinity of image fibers are identical to unaffected fibers. Bent, crushed, or otherwise damaged fibers show strain dichroism and will give an erroneous index.

Experiments scorching normal linen fibers agree with observations on scorched fibers from the Shroud. As the scorch color deepens, the two indexes of the linen approach the same apparent value. The index observed is the average of all of the orientations of the microcrystalline zones in the pseudomorph. Similar fibers have not been observed on image tapes.

Other than observing colored medullas, crystallinity and birefringence give good clues for differentiating between scorched and image fibers. The evidence is strong that the image is not a result of dehydration of the cellulose by any mechanism, thermal or radiation. ***The cellulose of the image has not changed as a result of image formation.***

A number of components for the development of a complex image-formation hypothesis are suggested by the bands of color on the cloth, the differences between image and scorched fibers, the interactions between the bands of color and the image areas, the density of the image in the vicinity of the nose and mouth, and the resolution of the body image (especially the fingers). These preliminary observations can be refined by using more "scientific" techniques.

Even if the chemistry of image formation could be explained in detail, the distribution of the color would still be a major question. Most of the image color appears only on the very topmost surface of the cloth, in discrete areas at the highest part of the weave.

Sam Pellicori of STURP had done image-formation experiments with lemon juice. He noticed that when color formed, it appeared mostly on the surface of the cloth.

A textile conservator in Turin, Anna Maria Donadoni, told us that after ancient cloth was washed in a *Saponaria* solution it was "laid out on bushes to dry." Under such conditions, materials that are in solution or suspension in the wash water will concentrate at the drying surface. This is a principle I have used to transfer traces of soluble materials from irregular surfaces into sheets of filter paper for later chemical analysis.

Evaporation concentration can explain the superficial nature of the image and the identical properties of the front and back images. An amine vapor that diffused from a body into the cloth could only react where impurities had concentrated. The amines do not react with cellulose.

The phenomenon can be demonstrated with a simple experiment. Prepare a dilute solution of food coloring, and divide it into two parts. Add a drop of liquid detergent to

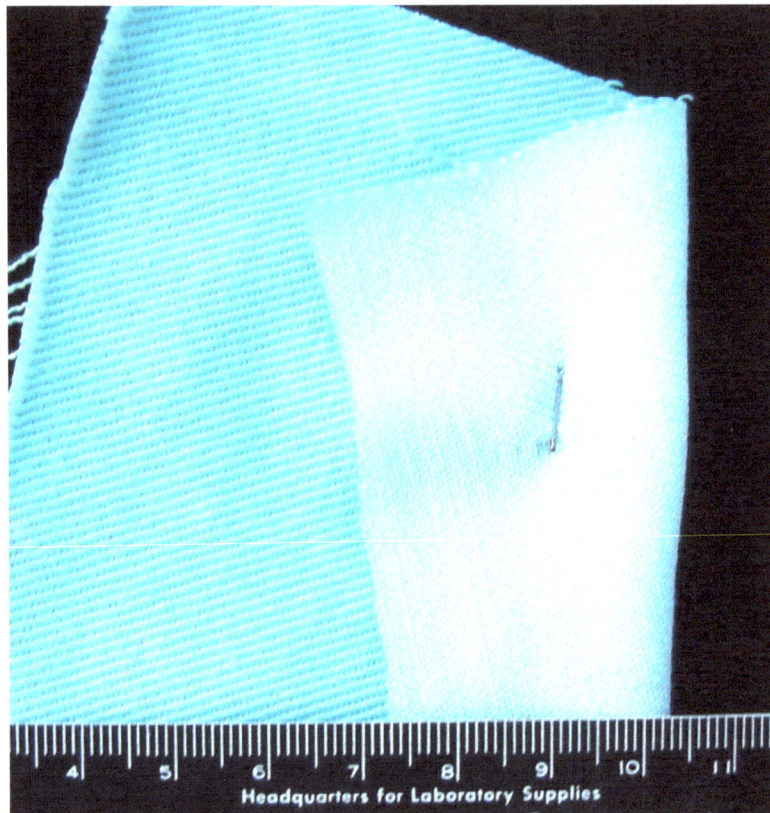

Figure VI-3: Demonstration of evaporation concentration made with modern, commercial linen and blue food dye.

one part. Cut some squares of white cloth that are about 10 cm on a side. Saturate cloth samples with one or the other of the solutions. Mark the samples for identification. Lay some saturated samples of cloth on smooth, non-absorbent surfaces (e.g., a sheet of plastic). Lay some samples on dry sand in the sun. Hang some samples from a line. Let the liquid evaporate. Different types of cloth will show different degrees of concentration of the dye on the evaporating surfaces, even on different adjoining fibers. It is possible to get dye concentration on *both surfaces*, while leaving the interior of the cloth white.

The puzzling "half-tone" effect has been mentioned. All of the colored image fibers showed approximately the same color density under a microscope. Assuming that the color formed by reactions with a very thin deposit of superficial impurities on the fibers, all of

34

the fibers should have shown identical spectra and roughly the same intensity of color. They did.

Another important observation is the fact that the image-forming process produced slightly different color densities (but identical spectra) on the different lots of yarn. We see this as the bands of different color in both the background and the image. The color-density of the image is not simply a function of the chemical properties of cellulose: It also depends on the individual properties of the batches of yarn. The observed effects must have been caused by different amounts of impurities that originally coated the surfaces of the different hanks of yarn as a result of slightly different bleaching conditions.

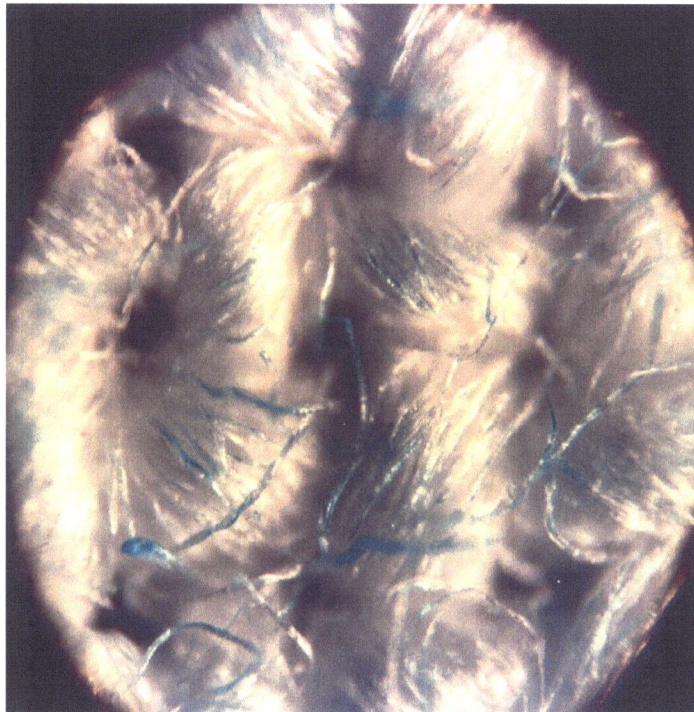

Figure IV-4: Photomicrograph (60X) of blue-dye concentration. The most deeply colored fibers are on the evaporating top of the weave.

Slightly different amounts of impurities on the different batches of linen yarn would cause slightly different surface energies. One major linen impurity is "flax wax," and it produces a hydrophobic surface. Liquids wet the threads as a function of the difference between the surface tension of the washing solution and the surface energy of the specific linen yarn. This would explain the "banded" appearance of the Shroud. The original observations and experiments on this phenomenon were done by Benjamin Franklin in 1774.

CHAPTER VII: CHEMICAL TESTS

The primary assignment given the chemical section of STURP was to investigate the claim that the image had been painted. The photographs available at the time did not give much detail on the yellow-brown color of the image; therefore, we had to consider all possible pigments and dyes. During and before the 14th Century, gold metal was the most important yellow. That would easily be detected by x-ray fluorescence. Other pigments in common use were yellow ocher (hydrated Fe_2O_3), burnt ocher (hematite Fe_2O_3) and other ochers, orpiment (As_2S_3), realgar (AsS), Naples Yellow ($Pb_3[SbO_4]$), massicot (PbO), and mosaic gold (SnS_2). Organic dyes included saffron, bile yellow, buckthorn, and weld. Madder root began appearing in Europe from the Near East about that time. Many of the dyes required mordants, which are hydrated oxides of several metals (e.g., aluminum, iron, and chromium). We decided that both chemical tests and x-ray fluorescence spectrometry would be required to confirm specific pigments.

Image color does not appear under the blood stains when they are removed with a proteolytic enzyme. Whatever process produced the image color must have occurred after the blood flowed onto the cloth, and the image-producing process did not destroy the blood. This indicates mild conditions for image formation.

On January 21st and 22nd, 1980, the members of STURP met at the Air Force Academy in Colorado Springs, CO. The chemical section had access to some good microscopes, and we had brought all of the critical reagents we would need to test for blood, blood serum, and different classes of colored materials.

On the basis of his work with microscopes, Walter McCrone had claimed that the image had been painted with glair (egg) and hematite. He had used an amido black reagent to detect the proteins. The amido black reagent had been developed to test paintings for the presence of glair, and it was not intended to be used on porous, adsorptive surfaces. Our tests showed that amido black did indeed give a positive test for proteins on linen: It also gave a positive test on clean, modern, commercial linen. That being the case, we studied it to determine the probability of false positive tests. The most definitive way to do this was to measure the R_f (rate of migration) of the reagent on cellulose surfaces. In performing the test, excess amido black is washed off of the sample with 5% acetic acid; therefore, we measured the R_f using a mobile phase of 5% acetic acid on different types of cellulosic materials with amido black. We found that the R_f was 0.05 ± 0.02: the amido black hardly moved! Further attempts to wash the amido black off of linen fibers showed that the dye was almost irreversibly adsorbed. It would be impossible to wash off of unpainted linen surfaces while following standard procedures. False positive results were inevitable.

McCrone had not followed the simplest procedures of rigorous analytical chemistry: He had not run "blanks." He did not test his method for false positives. All he wanted was to debunk the Shroud. A rigorous scientific study requires as many independent observational methods as possible. It is as unconscionable to allow antireligious sentiments to direct science as it is to demand a specific theological answer. I was disappointed to find that Walter could not be objective when he wanted publicity.

A more reliable test for proteins is the biuret-phenol test. We used a commercially prepared reagent from Hycel. We also used a home-made Fischer-Folin reagent for confirmation. When a protein is at least partially soluble, the intense blue-violet color of the reagent goes into solution around a fiber sample. We could see positive tests for proteins with the unaided eye as a blue spot grew around Shroud blood fibers on microscope slides.

Since all blood fractions contain sulfur in sulfoproteins, I also made an iodine-azide reagent. It is extremely sensitive, and it bubbles furiously with nitrogen gas in the presence of a sulfur compound (figure VII-1). This provided an independent microchemical test for proteins of the type found in blood.

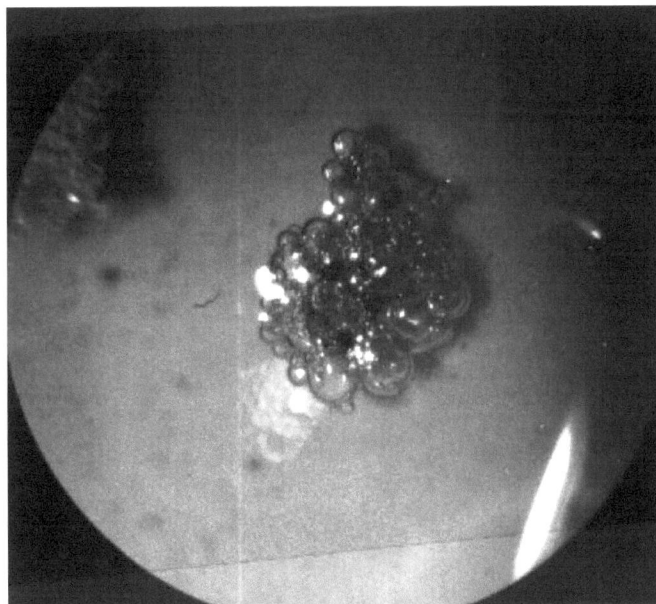

Figure VII-1: Iodine-azide test on microscopic blood fleck. Nitrogen bubbles prove the presence of sulfoproteins.

Joan Janney (now Joan Rogers) developed a method for recovering fibers from the tape samples that Walter McCrone had nearly destroyed. She glued the cover slips to a slide with a colorless epoxy cement and stripped the tape off of the cover slips. She then pulled individual fibers out of the adhesive with microforceps while watching with a microscope. She also excised small sections of tape with a scalpel. We then washed the fibers and excised tape samples with xylene to remove adhering adhesive. It was a laborious process, and nobody involved had much good to say about Walter.

We tested Shroud fibers and fibers from Raes threads that had been coated with a 10% egg white suspension and dried. All gave positive protein tests with all of the reagents. All fibers taken from sampling tapes that were pulled from blood areas gave positive protein tests.

Eric Jumper saturated a yellow fiber from sample 6BF with amido black. That sample was taken from a blood area that was near a scorch. It would have been expected to be coated with blood serum that had been denatured by heat and/or formaldehyde in 1532. The coated fiber did not give a protein test: All of the amido black washed off with 5% acetic acid, and it did not even produce bubbles in iodine-azide reagent.

I had preserved the envelope that presumably had contained the Zina thread. Using a low-power microscope, red flecks could be seen on the *outside* of the inner envelope as well as inside. The largest red fleck I could find, approximately 5 micrometers in diameter, tested positive for sulfoproteins with the iodine-azide reagent. Most of the fibers seen inside the envelope were cotton, obviously detached from the rag-bond paper while the sample was being picked to pieces. People who do not know what they are doing should do nothing around irreplaceable samples.

Adler and Heller later ran confirmatory tests with biuret/Lowry, coomassie blue, bromothymol blue, amido black, bromocresol green, and fluorescamine reagents. They included different kinds of controls, including samples that had been cleaned with a protease enzyme. Positive tests were obtained from fibers from blood-stain areas: Negative tests were always obtained from the yellow image fibers. They determined that the fluorescamine test could detect 1 to 10 *nano*grams of protein.

No positive tests for proteins were obtained with the reliable reagents applied to Shroud fibers from either background or pure image areas. *No proteins had been added to image areas.* The tests proved that the image was not painted with an egg-tempera system, as claimed by McCrone. The tests also make us question any contribution of body fluids to image formation.

When sulfoproteins are heated, they evolve hydrogen sulfide (H_2S), and they no longer give the iodine-azide test. McCrone had declared that the "Shroud blood can not be blood, because it doesn't give the iodine-azide test." The only sample he tested was 6AF, taken from scorched blood near the "lance thrust." It is easy to get the results you want if you are willing to throw out all conflicting evidence.

Microchemical spot tests with aqueous iodine indicated the presence of some starch fractions on Shroud fibers. This agrees with Pliny the Elder's description of linen technology. It would have been nearly impossible to wash all of the components of crude starch out of a cloth, even with modern detergents and methods. It would have been much less likely to wash the cloth completely clean with "struthium." A conclusive bit of evidence could have been obtained by the detection of any component of struthium, *Saponaria officinalis.*

Saponaria hydrolyzes to produce some aglycones that are fluorescent in the blue-white region, and the non-image part of the Shroud is weakly fluorescent with a maximum at about the correct wavelength, 435 nanometers. The fluorescence supports the idea that the cloth had been washed in struthium. The image filters the background fluorescence, making the image appear dark.

Saponaria is toxic, and it is a potent preservative. A textile conservator in Turin told us that old cloths that were washed in *S. officinalis* tend to be better preserved than newer ones. Comparison samples loaned to us by the Museum of Egyptology in Turin were still supple, and several dated to several thousand years BC.

Saponaria produces four glycosidic saponins, all containing gypsogenin. Gypsogenin is built on a triterpene ring structure, and cloths washed in *S. officinalis* fluoresce as they hydrolyze and age. We did not have the equipment to test the fluorescence spectrum in 1977, but it seemed to be in the blue-white region. I believe that washing in *S. officinalis* could explain the background fluorescence of the Shroud.

The glycosides hydrolyze to produce sugar chains. The following carbohydrates were identified in those chains: galactose, glucose, arabinose, xylose, fucose, rhamnose, and glucuronic acid.

Pentose sugars with a furanose structure appear to be the most reactive sugars. I have used them for chemical kinetics experiments. The *Saponaria* sugars should be quite chemically active. Detection of sugars that were derived from *S. officinalis* would provide strong evidence that ancient technology had been used to prepare the cloth used for the Shroud.

In order to make a more detailed analysis for possible flax impurities and/or sugars from *Saponaria officinalis* (the "struthium" mentioned by Pliny the Elder), I made some Bial's reagent (orcinol, con. HCl and $FeCl_3$). It gives a bright Kelly green color with pentose sugars or furfural. I could not get a clear positive test for pentoses from Shroud samples; however, I got some fairly weak tests for pentoses from Raes threads. That should have been a clue. The Raes threads showed different chemical characteristics than the main part of the Shroud, but I ignored the fact until much later. It is important in a discussion of the sample taken for radiocarbon analysis in 1988.

If the image had been formed by a scorching-type, high-temperature reaction, some pyrolysis products of linen, including furfural, might still be present. The detection of pyrolysis products would have been fairly conclusive evidence for an image-formation mechanism; however, the absence of such products would prove nothing. I got no test with Bial's reagent, so I also tried Seliwanoff's test for furfural. It gives a nice, bright red color with furfural, but it gave no test with fibers from a light Shroud scorch. Furfural polymerizes over time to form a dense, dark polymer that does not give the test. Polymerization is faster when the reaction is catalyzed with some common impurities, and it can be slowed with inhibitors. I could not prove the presence of furfural on image areas; however, it was worth the effort to try. The same tests can detect pentose sugars.

I could not prove the presence of pentose sugars on the Shroud, so I could not prove that the cloth had been washed with *S. officinalis*. Only the fluorescence evidence remains to suggest the use of struthium.

The Seliwanoff's reagent also gives a red color with levulose (fruit sugar), but it does not react with levulinic acid (a cellulose pyrolysis product). I got a red test with scorched Shroud fibers, but background fibers gave no color. I suspect that the literature descriptions of the reagent are not complete.

The photomicrograph of image fibers in Chapter VI shows dark deposits at the growth nodes of the linen. I assumed that these spots were lignin that was not removed during the bleaching process. Modern linen that has been bleached with chlorine or other active bleaches shows some very small black specks at growth nodes. I thought that an abundance of lignin would give evidence for primitive technology.

A very sensitive test for lignin can be found in the scientific literature. It uses phloroglucinol in concentrated hydrochloric acid to produce and react with vanillin from the lignin. The positive response is a vivid violet color. The Shroud fibers did not give the test. The small specks of black on modern linen did not give the test; however, the black deposits at the growth nodes on fibers from the Shroud's medieval backing cloth (the "Holland cloth") showed clear positive tests. Other medieval samples we had gave a clear test. A sample from the wrappings of the Dead Sea scrolls did not give the test, but that is not a place I would want to live without air conditioning.

Dr. Stan Kosiewicz of Los Alamos, an analytical chemist, aged samples that contained lignin at 40, 70, and 100°C for up to 24 months. Just as we suspected, he found that lignin loses its vanillin with heating and/or time. You can smell the vanillin coming off of some kinds of tree bark, which are rich in lignin. Smell a Jeffrey pine in the sun some time. Stan found that the upper limit of the experiment was about seven months at 100°C, because the cellulose began to turn brown.

Chemical rates depend very strongly on temperature. Rates of all kinds of reactions are modeled with an exponential equation called the Arrhenius expression:

$$\frac{d\alpha}{dt} = kf(\alpha)Ze^{-E/RT}$$

Rates can be predicted from amounts of reactants (α is the fraction reacted at any time t) and known, measured chemical kinetics constants (k, the rate constant; Z, the Arrhenius frequency factor; E, the Arrhenius activation energy; R, the gas constant; and T, the absolute temperature in degrees Kelvin). Any chemical process involved in Shroud aging or image formation will have properties in accordance with this equation. Notice that the rate of any chemical reaction will not become zero until the temperature reaches absolute zero, but some things take a lot of time. The temperature appears in the exponential; therefore, it is usually more important than amounts of reactants; however, when all reactant has been depleted, the rate approaches zero. The f(α) is called the "depletion factor," and it depends on the physical state of the reactant, the type of reaction, and/or the number of molecules involved in the reaction. Both temperature and amount of reactant available are important.

The rates of lignin degradation were so slow that we could not make accurate determinations of the rate constants in the amount of time we could spare from making an honest living, but we could make estimates. We used the time until the phloroglucinol/HCl failed to detect lignin as the criterion at 40, 70, and 100°C, not a very rigorous method; however, it gave the following Arrhenius predictive model.

$$k = 3.7x10^{11} e^{-\frac{29,600}{1.987T}}$$

So roughly how long would it take vanillin to disappear from Shroud fibers? The real answer in terms of chemical kinetics is: it won't ever all go away. The reason is that the rate law is exponential: $-\ln(1 - α) = kt$ (where α is the fraction of the lignin that has lost its vanillin at any time t). But the lignin test has its limits. Using the detection limit as the criterion for "no vanillin," we can make some estimates for the real-world time it would take for the lignin's vanillin to disappear.

It is interesting to compare the numbers for lignin with those that have been determined for cellulose and some other saccharides. For example, the kinetics constants for the first dehydration reaction of cellulose are about 47,300 cal/mole and its pre-exponential is about 3.2×10^{14} s^{-1}. The constants for glucose, the most common sugar, are E = 23,900 cal/mole and Z = 1.26×10^{10} s^{-1}. The activation energy for arabinose (a less-stable pentose sugar) is about 19,700 cal/mole.

The high activation energy for cellulose reflects the fact that it is a crystalline solid. It has long been known that the activation energy for a pure, crystalline material is close to the sum of its latent heat of fusion (how much heat it takes to melt the solid) and the activation energy of the compound in a liquid phase. Cellulose is very stable when it is solid, but it melts with decomposition when it is heated quickly to 260-270°C. Melting with decomposition is an autocatalytic process, but it stops completely when the liquid cools and freezes.

Cellulose starts to color pretty fast at about 325°C. That is probably close to the temperature at the margins of the scorches during the fire of 1532. At that temperature, it would take about 0.6 second to decompose 95% of the lignin, but lignin decomposes very slowly at normal temperatures.

Since the rate law is exponential, the maximum diurnal temperature is much more important than is the lowest storage temperature; however, if the Shroud had been stored at a constant 25°C, it would have taken about 1,319 years to lose 95% of its vanillin. At 23°C, it would have taken about 1,845 years. At 20°C, it would take about 3,095 years. At 15°C it would take about 7,482 years, and at 10°C it would take about 18,661 years. You can see the problem: time is sensitive to the temperature history.

We were talking about a 95% loss of vanillin, but how about cellulose? According to the published numbers, it would take about 84.3 *billion* years to decompose *1%* of it under dry, sterile conditions. That is about **six times** the age of the universe. We do not need to worry about cellulose decomposition in the Shroud, unless someone contaminates it.

I have used this same method to estimate the maximum temperatures seen by nuclear weapons that had been in storage in known locations, and the chemical estimates checked with measured temperature/time values. Unfortunately, the time and money are not available to study the Shroud as carefully as a weapon. The Shroud can make you reflect on the priorities of humanity.

If the Shroud had been produced between AD 1260 and 1390, as indicated by the radiocarbon analyses, its lignin should be easy to detect. A linen produced in AD 1260 should have retained 37% of its vanillin in 1978. All of the medieval linens we have tested gave a good test for vanillin.

The thermal conductivity of linen is very low, 2.1×10^{-4} cal cm^{-1} s^{-1} °C^{-1}, so the unscorched parts of the folded cloth could not have gotten very hot in 1532. The temperature gradient through the cloth in the reliquary should have been very steep, and the cloth's center would not have heated at all in the time available. The rapid change in color at scorch margins from black to white illustrates this fact.

If you really want to calculate the temperature profile through the cloth, all you need is the dimensions of the box, the thermal conductivity of the cloth at its packing density in the box, the heat capacity of the cloth, the kinetics constants, and the endothermic heat of reaction. Assuming that there is no endothermic reaction, a situation that would obtain only in the parts of the cloth where we see no scorches today, you can make a one-dimensional calculation for the exponential temperature profile from the following equation:

$$T(t) = T_0 + (T_\infty - T_0)\left[1 - e^{-t/\tau}\right]$$

where t is time in seconds; T_0 is the initial temperature; the surface temperature when heating was stopped is T_∞; and τ is the time constant. The time constant is calculated from the following equation:

$$\tau = a^2/k$$

where a is a dimension and k is the thermal diffusivity in cm^2/s. If the shape is anything other than a perfect sphere, an infinite cylinder, or an infinite slab, you run into more complicated calculations.

If you want to make your calculations inside zones where the cloth is smoldering or decomposing, you must use equations suitable for reactive heat flow. At that point, it would be useful to consult the following sources: 1) D. A. Frank-Kamenetskii, *Diffusion and Heat Transfer in Chemical Kinetics*, Plenum Press, New York, NY (1969). 2) N. N. Semenov, *Chemical Kinetics and Chain Reactions*, Oxford University Press, London (1935).

Sometimes it is simpler to do an experiment. The fire of 1532 did that experiment for us. Any heating at the time of the fire would decrease the amount of vanillin in the lignin as a function of the temperature and time heated. Different amounts of vanillin would have been lost in different areas, depending on the temperature and how long the temperature took to return to normal.. No samples from any location on the Shroud gave the vanillin test. Since the Shroud and other very old linens do not give the vanillin test, I believe that we can be confident that the cloth is quite old. It is very unlikely that the linen was produced during medieval times. See Chapter IX for an explanation of the difference between the radiocarbon age determination and the observations on the linen's production technology and composition.

Another, more-qualitative observation can be made on the scorches: their margins are very smooth. There is no evidence for an uneven distribution of catalysts on the cloth.

With enough work, we could get more accurate and confirmed kinetics numbers for lignin. It would be a good thesis project for some chemistry student who is interested in the Shroud. However, under any assumptions we may make, chemistry would suggest that the cloth is older than the published radiocarbon date. There are other reasons.

A corollary of the lignin problem is that all degradations will have the same exponential dependence on temperature. If you want to preserve the Shroud, cool it down. But don't cool it too much. A hard freeze would probably fracture fibers.

Biblical accounts suggest that aloes and myrrh were used as Jesus' body was prepared for burial. It was relatively simple to test for the presence of both of these materials. Dr. Baima Bollone, a professor of forensic medicine in Turin, had claimed to have found aloes and myrrh on the shroud by antibody-antigen testing, but he did not publish details or ask for confirmation. These claims needed confirmation by independent observations.

Aloes contain a complex mixture of chemical compounds, but all contain a fused-ring structure called anthraquinone that is made of three benzene rings with double-bonded oxygens attached to the middle ring. For example, aloin contains an anthraquinone structure with a pentose sugar attached. Aloe emodin has two -OH groups and no sugar, and it turns pink in strong ammonia water. Aloe emodin anthranol has -OH groups where the doubly-bonded oxygens usually are, and it fluoresces in the green. The Shroud fluoresces in the blue. All of the aloes fractions have extremely intense spectral characteristics, and are easy to detect by reflectance spectrometry.

Myrrh has a piercing odor, and it is a yellow to yellowish-green sticky, thick material. It contains 2.5 - 8% of a volatile oil that is about 1% cuminol. The rest is eugenol, metacresol, pinene, limonene, dipentene, and two sesquiterpenes which are all compounds with known properties. It contains about 25 - 40% resin, which is composed of resin acids ($\alpha-$, $\beta-$, and $\gamma-$commiphoric acids), resenes, and phenolic compounds. Testing converts one phenolic compound into protocatechuic acid, which gives an intense violet color with ferrous ions. Another gives pyrocatechol, which takes part in many sensitive microchemical color reactions. Myrrh also contains about 60% of a gum that is much like acacia gum, giving arabinose sugar on hydrolysis. Several sensitive methods should be able to detect myrrh. We could not confirm Bollone's claim for either aloes or myrrh.

On 24 March 1979, STURP met in Santa Barbara, California. Preliminary presentations were made on all of the chemical, microscopic, and instrumental analyses. Two important claims were made by Walter McCrone. He said that *all* of the red flecks in blood areas are iron oxides in different degrees of hydration. One of his own people later refuted this during laser-microprobe analyses. Walter also stated that he had found wheat starch on the Shroud. We confirmed this by microchemical testing with aqueous iodine, supporting an hypothesis that the cloth had been made by ancient methods.

On 14 March 1981, STURP met in New London, Connecticut, to report on later studies. The most surprising results were reported by Professor Alan Adler of Western Connecticut University. He had found that the image color could be reduced with a diimide reagent, leaving colorless, undamaged linen fibers behind. This confirmed spectral data that indicated that the image color was a result of complex conjugated double bonds. He made the reagent with 70% hydrogen peroxide (H_2O_2) plus 97% hydrazine (N_2H_2) in boiling pyridine. *Caution: This reagent can be dangerous to handle.*

Up until this time, we had assumed that the image color was a result of chemical changes in the cellulose of the linen. The most likely change would involve the dehydration of the cellulose to produce conjugated-double-bond systems Adler's observations strongly suggested that the cellulose was not involved in image formation. *This is an extremely important observation.*

We could hardly believe this observation, so Adler went back to his laboratory to make more detailed observations on the tape samples. He immediately reported that he had seen "ghosts" in the adhesive of the tapes. It appeared that some image fibers had been pulled out of the adhesive, *and their colored coating had been stripped off of the fiber and remained in the adhesive.* He found that the colored "ghosts" showed the same chemical properties as the authentic image fibers from the Shroud. I went to his laboratory to confirm his observations, and he certainly was correct. The observation was so important that I made another confirmation by looking for as many ghosts as I could find on slides in my laboratory. They appeared on all image tapes (figure VII-2 is one example).

The bands of color (Chapter VI), and the fact that all of the image color appears only on the outer surfaces of the fibers, suggested that image formation involved a thin layer of impurities. Since the cellulose was not colored, the impurities had to be significantly less stable than cellulose. This also suggested that the impurities were the result of cloth-production methods (Chapter IV), and they should appear on all parts of the cloth. A search for carbohydrate impurities on the Shroud confirmed McCrone's detection of some starch fractions. Starch and low-molecular-weight carbohydrates from crude starch would color much more easily than would cellulose as a result of either thermal dehydration or chemical reactions.

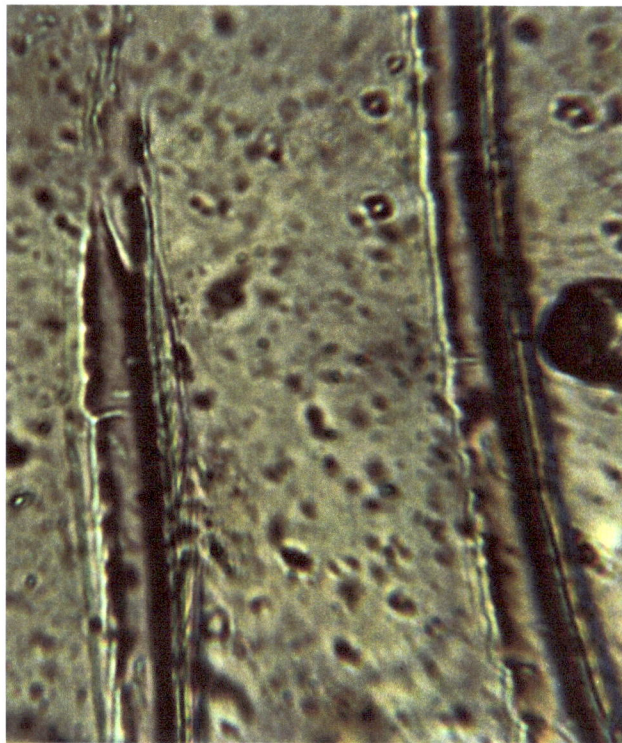

Figure VII-2: "Ghosts" of colored image fibers that remained in the adhesive of sampling tapes after the tape was pulled from the Shroud's surface (800X).

If preexisting impurities enabled image formation, some should have still been on the Shroud at the time of the 1532 fire. A search of tape samples from lightly-scorched areas revealed ghosts that appeared to be identical to those from image areas. Thin layers of colored impurities had stripped off from scorched fibers that were completely isolated from image areas (figure VII-3).

Scorched fibers from the sample shown in the figure (STURP sample 1IB) were very slightly colored; however, scorches on the Shroud ranged from almost invisible to black. All degrees of cellulose damage have been observed between uncolored and carbonized black. More heavily scorched areas show darkening of the cellulose itself, and many of the medullas appear to be darker than the body of the fiber (figure VI-1).

Chemical tests seemed to give strong support for rejecting any painting hypotheses for image formation. The most surprising outcome of the chemical tests was the fact that image color appeared *only* on the outer surfaces of colored image fibers. This required a major rethinking of image-formation hypotheses. Any image-formation mechanism that would result in color formation *inside* the linen fibers must be rejected. Some "theories" that have been mentioned that would cause coloration inside fibers are penetrating radiation, high temperature scorching (hot statue, painting with a torch, etc.), and catalyzed dehydration of the cellulose. *Image fibers are colored only on their surfaces.*

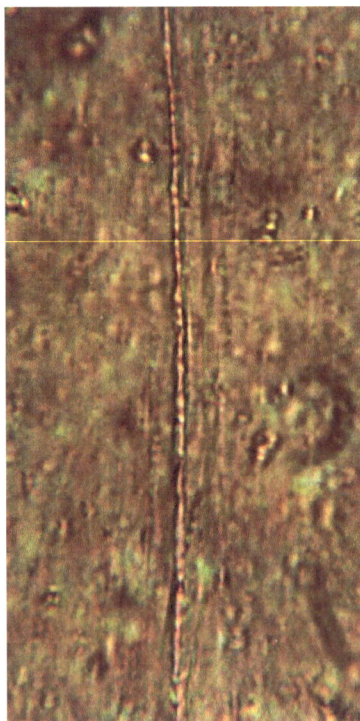

Figure VII-3: A line of yellow flakes stripped off of one side of a lightly-scorched fiber (800X). The outline of the other side of the fiber and some dispersed flakes are visible.

CHAPTER VIII: INSTRUMENTAL METHODS OF ANALYSIS

People are very clever. It was an interesting challenge to try to guess all possible ways an image could be painted or hoaxed. Because we might not think of them all, we tried to make sufficient observations on the cloth and all associated materials to detect unexpected things.

If the shroud was not a hoax, we might find evidence that it had been in contact with a body. It would be very useful to prove the cloth was actually a shroud. The same observational methods used to detect fraud could be used to detect the components of sebum and impurities. However, clever people might use an animal to produce a hoax. We had to plan to look for sebaceous excretions from all kinds of animals. The list we used is shown in the table.

SEBACEOUS EXCRETIONS OF HOMO SAPIENS AND SEVERAL ANIMAL SPECIES (%)								
CONSTITUENT	H. sapiens	Sheep	Guinea pig	Mouse	Rabbit	Rat	Ox	Duck
Free fatty acids	28.3	11.0	6.0	7.5	9.0	7.4	5.1	
Combined fatty acids (esters, waxes, triglycerides)	34.6	44.0	49.3	36.7	43.6	51.1	55.4	17.6
Triglycerides	32.5	0	0	0	0	0		
Unsaponifiable matter	30.1	46.1	44.8	51.6	45.9	41.1	42.7	
Squalene (very fluorescent)	5.5	0	0	0	0	0		
Hydrocarbons	8.1	1	1.5	1.1	3.5	1.5		
Wax alcohols	6.2	9.0	5.0	5.9	31.0	17.5		48.0
Cholesterol	4.1	10.0	18.0	4.5	3.5	5.8	14.4	1.4
Dihydrocholesterol	0.1	2.5						
Lathosterol			1.8	8.1	0.1	4.4		
Other sterols	0.3	1	0.7		0.3	0.2		
Isocholesterol	tr	12.5						
Alkane-1,2-diols	2.0	2.3	5.6	27.5	2.2	2.9		

The most stable constituent of sebum is squalene, it has a very high boiling point (280°C at 17 mm pressure), and it is also unique to human beings. It is very fluorescent, and it might be detected and identified by its spectrum. It has a mass of 411, and it should be possible to detect it with great sensitivity using mass spectrometry. It would prove contact with a human body.

The plans for analyzing the Shroud were based on the requirements for detecting and identifying all of the types of compounds that would be diagnostic in determining the "authenticity" (whatever that is taken to mean), age, and composition of the cloth and image.

The Shroud was observed by visible and ultraviolet spectrometry, infrared spectrometry, x-ray fluorescence spectrometry, and thermography. Later observations were made by pyrolysis-mass-spectrometry and laser-microprobe Raman analyses.

Before looking for evidence of painting, it was important to make a literature survey for all of the types of materials that had been used through the ages.

The oldest known paintings appeared in prehistoric times, and they are found in the caves of Europe and Africa. They were done in natural materials, e.g., red and yellow ochre and charcoal. There is evidence that the pigments were mixed with animal fat for application to the irregular cave surfaces. Encaustic appeared very early in history, applying pigments in wax that was then heated to fusion.

The fresco and mosaic methods clearly could not apply to the Shroud, but we considered them in the search for unexpected materials.

Tempera painting was used by the ancient Egyptians and Greeks, and it was perfected by the icon painters between about AD 400 and 1200. It was essentially the only painting method used during the Renaissance until the 15th Century. It involves powdered pigments mixed with many different water-soluble binders. The primary binders were egg white, egg yolk, plant gums (especially gum Arabic), resin (e.g., Chios mastic), wine, honey, and/or animal glues. Tempera would have been the most probable painting method for perpetrating a hoax with the Shroud at the time indicated by the 1988 radiocarbon analysis, AD 1260-1390.

Over 40 different pigments were used in medieval times. Some were raw minerals ground by the artist or obtained from traveling dye-and-gem merchants. The determination of particle-size distributions would help detect man-made colored particles.

Some pigments were manufactured by alchemists, requiring elaborate (and toxic) processes. Much could be learned by observing the properties of such materials. And it would be interesting to observe reactions between the pigments and the painting media. Many man-made pigments react with the medium over some length of time.

Pure colors were laid on, then shading was done with transparent layers of black; therefore, we might expect to see variations in elemental composition with location on a painted image.

Blues were obtained from ultramarine blue, which was produced by a complex refining process of lapis lazuli. Azurite, malachite, and indigo (extracted from woad and usually used as a lake).

Red pigments included lead tetroxide and vermilion (red mercuric sulfide), both produced by alchemists. Hematite, red clays, Sinopia, Venetian Red, and Terra Rosa were also used. Dragonsblood is the resin of a shrub in Eastern India. Brazilwood came from trees in Ceylon. It was filed to a fine powder, boiled, and mordanted with lye for a purple color and alum for an orange-red.

Yellow was usually produced with ochers, which came from natural deposits of hydrated iron oxides and clay minerals. Orpiment is arsenic sulfide, and it reacts with other pigments like white and red lead and verdigris. It turns black, unless tempered with glue. It would be interesting to find on the surface of the Shroud. Realgar, orange arsenic sulfide,

48

is much like orpiment; however, it is even less stable. Gamboge, Spanish yellow, appeared from the East during the Crusades. It is the sap of a tree, but it must be refined properly for use. Massicot, lead monoxide, has been used since Classical times. Gallorino, yellow lead oxide, often is found around volcanoes. Alchemists made it by roasting white lead. Arzica is an Italian dye made from weld. Saffron is made from the stamens of the crocus. We considered it a possible stain for producing the image. Mosaic gold is yellow tin sulfide, and it was used to imitate gold. Celandine, aloes, bile yellow, and shell gold (gold-leaf powder) were also used to produce yellows.

We considered that white pigments could be used as an overlay to control shading, or they could appear as impurities in a painter's environment. The most probable ones would be zinc oxide and white lead (which rapidly changes color). Blacks might also be used to control shading, and most were obtained from burned vines and lamp black. There were even reports of squid ink being used.

We thought that the most probable pigments, given the reported color of the image as of 1977, would be hematite or saffron with carbon shading.

It would be extremely unlikely that oil paints had been used to hoax the image during or before the 14th Century; however, we planned observations that would detect such materials. Oils were the favorite vehicles for pigments during the time of the 1532 fire. They could have been used in an attempt to reproduce the Shroud, if it had been totally destroyed in the fire.

Reflectance Spectrophotometry:
Your eye sees some materials as colored, because the surface absorbs some wavelengths of light and reflects others. A red surface absorbs all visible wavelengths other than red. Each chemical compound absorbs wavelengths that are characteristic of its chemical structure. An opaque, solid surface poses special problems for analysis, especially when you are not allowed to destroy any of the material. It was necessary to observe the spectra reflected from the solid surface of the Shroud in order to obtain chemical information.

Roger and Marty Gilbert of Oriel Optical Corp. designed and built a dual-beam reflectance spectrometer that could be used on the Shroud. They recorded visible, ultraviolet, and fluorescence spectra from many areas of the cloth.

We were worried that my tape samples might damage the surface, especially in image areas, so the Gilberts took spectra of specific locations and marked the locations with small magnets. I took a tape sample at that location, and the Gilberts repeated their spectral analysis in the identical location. Because the before and after spectra were identical, we felt confident that we were not damaging the Shroud or contaminating the surface.

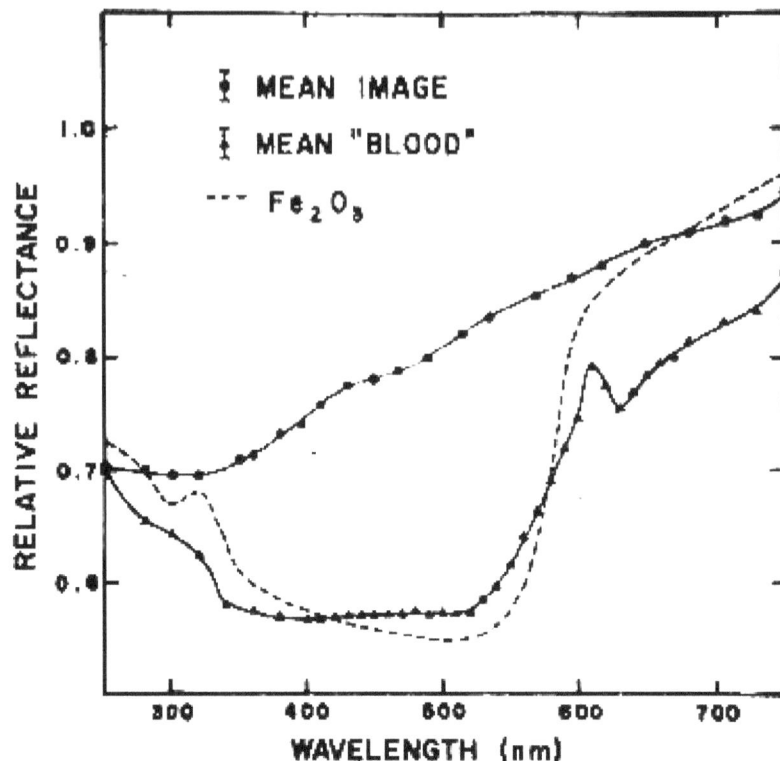

Figure VIII-1: Reflectance spectra in the visible range for the image, blood, and hematite. The image could not have been painted with hematite.

They found that the visible spectrum of the image, the quantitative analysis of the color, was identical to the color of old, sepia linen (figure VIII-1). This proved that the same general types of chemical structures were involved in the aging process of the cloth and the image color. The structures contained many conjugated double bonds. Adler's observations that the image color could be reduced and decolorized with strong chemical reductants confirmed this spectral observation.

They also found that hematite (Fe_2O_3), claimed by McCrone to have been used to paint the image, could not have contributed the color (figure VIII-1). Hematite absorbs light in the blue and green, reflecting the red. Your eye sees red, and the spectrum shows a sharp cutoff. Reflectance goes from nearly zero at wavelengths below about 550 nanometers to nearly 90% above 600 nanometers (in the red). All known pigments could be rejected in the same way as hematite. The image was not painted, unless it was painted with a flame that scorched the linen (that hypothesis will be discussed). The lack of detectable pigments agreed with the x-ray-fluorescence analysis and chemical tests.

50

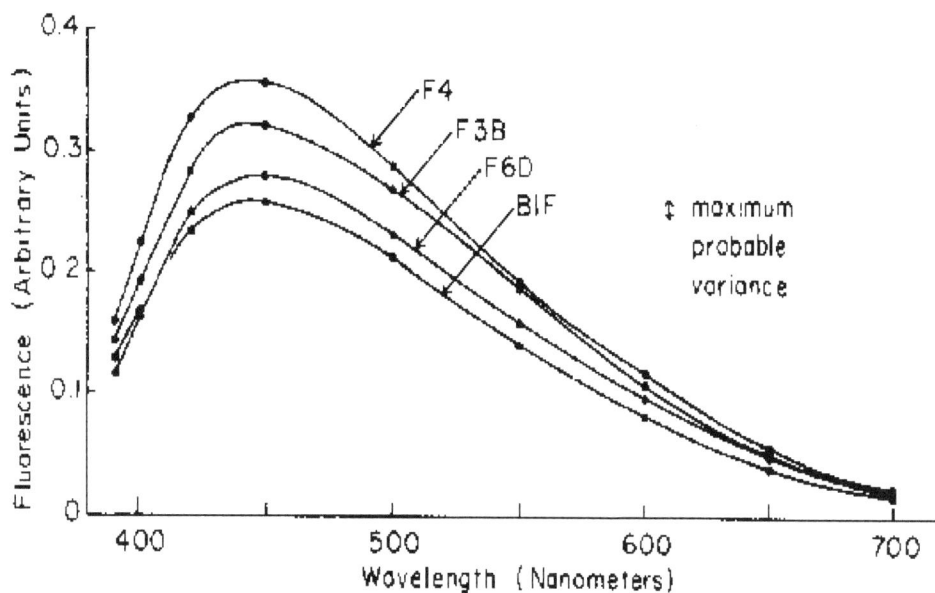

Figure VIII-2: Spectral fluorescence of four clear areas of the Shroud with excitation at 365 nanometers. Maximum fluorescence is at about 435 nanometers.

The background of the Shroud is weakly fluorescent with a maximum intensity at about 435 nanometers wavelength, in the blue. The image did not fluoresce at all. The background fluorescence was in the correct range to be explained by polynuclear aromatic chemical compounds, which could help confirm the technology used to produce the cloth. Some materials with the correct properties are produced by *Saponaria officinalis*, the "soapweed" that probably was used to wash the cloth after it was woven.

Sam Pellicori designed and built an independent spectrometer, and he confirmed the observations made by the Gilberts. The data enable some definite statements to be made: 1) The image was not painted with hematite, at least to a limit of five micrograms per square centimeter. 2) The image was not painted with any fluorescent pigments or media. 3) The color of the image is the result of conjugated-double-bond chemical structures.

The evidence from spectrophotometry was among the most important assembled during the studies of 1978.

Pyrolysis Mass Spectrometry:
Our primary goal in undertaking pyrolysis-MS analyses on samples from the Shroud was the sensitive detection of impurities (e.g., painting materials and sebum). Most of the structural materials and probable impurities in Shroud samples were carbohydrates. We wanted to see traces of materials that were not carbohydrates. That can be hard to do, but modern instrumentation enables several approaches to be taken.

As discussed in the context of the fire of 1532, materials that are heated in the absence of air (oxygen) tend to produce pyrolysis products that are often characteristic of the sample. These products can be produced in the inlet of a mass spectrometer (MS), and they can be separated and identified.

A MS puts an electrical charge on the products ("ionizes" them) and accelerates them through magnetic and/or electrically charged sectors. The products take different paths, depending on the mass of the molecule. Our samples were run at the Midwest Center for Mass Spectrometry (MCMS), University of Nebraska-Lincoln. This is a National Science Foundation "Center of Excellence," and it ranks among the foremost facilities in the world.

The methods used to ionize the products are important in determining the sensitivity of the method, and we were anxious to get maximum sensitivity in looking for trace impurities. Molecules in a high vacuum can be ionized by electron impact or chemical ionization. Chemical ionization uses collisions with excited atoms or molecules to ionize the sample, and it gives a much simpler mass spectrum than electron impact. Simple means that the molecules are not knocked to pieces in the ionization process. More ions with the maximum mass are produced, and more ions means higher sensitivity. The MCMS had the most sensitive MS at the time we were analyzing samples.

Walter McCrone had ignored agreements on how the STURP samples were to be observed, and he contaminated all of our tape samples by sticking them to microscope slides. All of the fibers were immersed in the tape's adhesive. Joan Rogers laboriously cleaned and prepared Shroud fibers for analysis at the MCMS. Washing them with solvents was certain to contaminate the samples with traces of the solvents, but we knew the molecular weights of the solvents. More irritating was the fact that washing was guaranteed to remove some of the impurities. The tape-sampling method had been planned such that all such contamination problems would be avoided. McCrone did not care.

The MCMS machine was sufficiently sensitive to detect parts-per-billion traces of the low-molecular-weight fractions (oligomers) of the polyethylene bag that Gonella had used to wrap the Raes threads. It did not detect any unexpected pyrolysis fragments that indicated any Shroud materials other than carbohydrates. That is exactly what would be expected from a piece of pure linen. This helped confirm the fact that the image was not painted.

Although McCrone had probably ruined our chances of finding squalene and triglycerides by pyrolysis/MS, we still had the results from the uv/visible spectra. When dealing with people, it is a good idea to have a back-up plan.

The matrix of samples that were analyzed at the MCMS is shown in the table. We attempted to get representatives of all of the different areas of the Shroud.

Slide Number	Image	Scorch	Blood	Water
6BF	yes	light scorch	Yes	no
3EF	yes	no	yes	no
Zina heel	yes	no	yes	no
Raes #3	no	no	no	no
1EB	yes	no	no	no
2CF	yes	no	no	yes
Edgerton modern	no	no	no	no

Materials that are heated in the absence of air (oxygen) tend to produce pyrolysis products that are characteristic of the sample. Organic materials pyrolyzed in air tend to produce heat, light, carbon monoxide and/or carbon dioxide, and water. It was fortunate that the Shroud was stored in a closed reliquary in 1532. The lack of oxygen in the heated reliquary made the intersections of scorches with image, blood, and water-stains important for a chemical study of the cloth. It is most unfortunate that no chemists were involved in the "restoration" of 2002, because most of the pyrolysis products were removed from the Shroud. For example, we will never be able to test light scorches for saffron products. The amount of chemical information lost is incalculable, and it illustrates the dangers inherent in "secret" operations.

Materials can also be heated to decomposition under vacuum in the inlet of a mass spectrometer, and they can be separated and identified. The technique is called pyrolysis/mass spectrometry. The MCMS pyrolysis system covered a wide range of temperatures, and most organic materials would have been detected.

The pyrolysis-MS analyses did not detect any nitrogen-containing contaminants in pure image areas. This seemed to rule out glair (egg white) as well as any significant microbiological deposits, confirming the microchemical tests that were also made (Chapter VII). They did not detect any of the sulfide pigments that were used in antiquity, e.g., orpiment, realgar, mosaic gold, and cinnabar (vermilion, mercury sulfide, HgS). The Shroud's image had not been painted with any known vehicles and pigments.

Many of the pyrolysis fragments observed by pyrolysis/mass spectrometry would be the same whether they came from cellulose, hexose sugars, pentose sugars, or starches. However, the ratios of products can be characteristic and important.

The major products of the thermal decomposition of cellulose and other carbohydrates in the absence of oxygen are water, formaldehyde, carbon monoxide, hydroxymethylfurfural (5-hydroxymethyl-2-furaldehyde, obtained only from hexoses), furfural (2-furaldehyde, a major product from pentoses), levulinic acid (4-oxopentanoic acid), and 3-pentenoic-γ-anhydride. Pentoses (sugars composed of five-carbon rings) do not produce any hydroxymethylfurfural. Hexoses, as in the cellulose structure (six-carbon glucose), produce hydroxymethylfurfural. The hydroxymethylfurfural deformylates (loses a formaldehyde fragment) in a series reaction, producing some furfural. Therefore, a pure hexose system will

53

show some furfural, but a pure pentose system will not show any hydroxymethylfurfural. For example, figure VIII-3 shows a mass spectrum that was taken when the first decomposition products started to appear over a sample of image fibers. The sample appears to be nearly pure polyhexose material, e.g., cellulose. There is a significant peak at m/e 126 (hydroxymethylfurfural), but there is not much at m/e 96 (furfural). The Shroud is nearly pure linen, but linen is not pure cellulose like cotton.

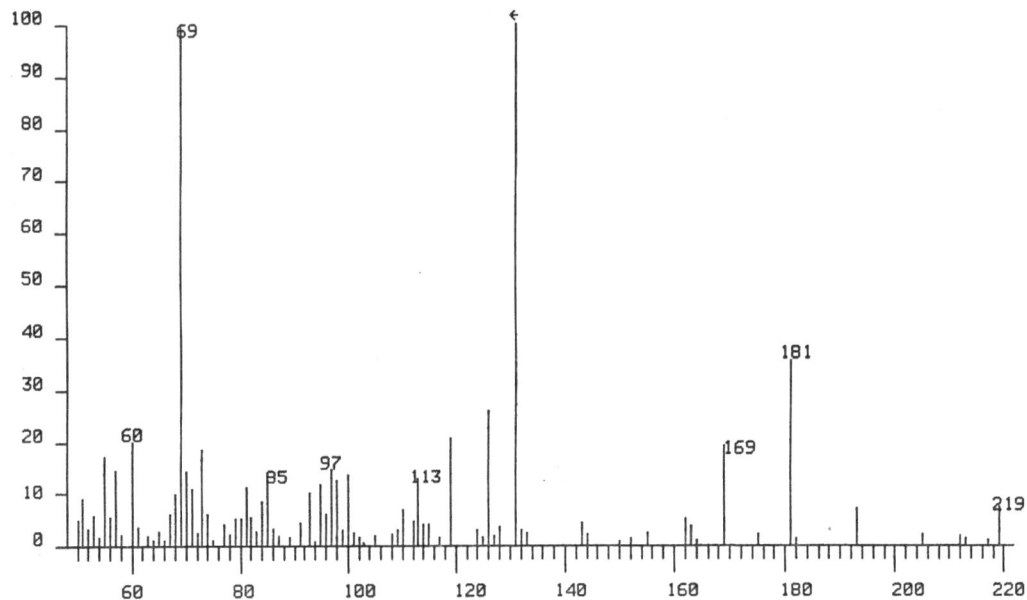

Figure VIII-3: Mass spectrum obtained from the pyrolysis of fibers from Shroud image sample 1EB. The sample did not contain any pentose sugars.

Figure VIII-4 shows that the low-temperature decomposition products obtained from the Raes sample (cut in 1973 from the area adjoining the radiocarbon sample of 1988) gave absolutely no m/e 126 signal: the cellulose of the sample had not yet started to pyrolyze. There is, however, a significant m/e 96 signal: furfural was being produced at this temperature. This proves that the sample contained some pentose-sugar units. This is unique among all of the Shroud samples: no other area showed this pentose signal. This would have been inexplicable but for the observations on the radiocarbon sample reported in Chapter IX.

The ordinate of each graph shows the relative ion intensity for each product produced at that temperature. The abscissa shows the mass of the ion.

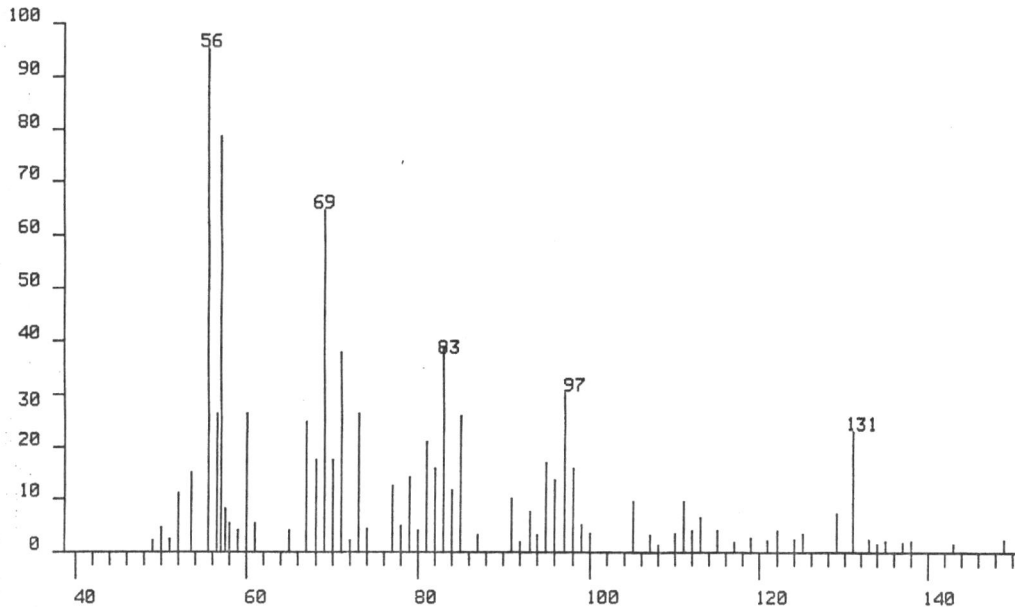

Figure VIII-4: Mass spectrum of the low-temperature pyrolysis of fibers from Raes sample #3. The peak at mass 96 shows that the sample contained a significant amount of pentose sugars.

In addition to different products, the different carbohydrates produce pyrolysis products at very different rates. Rates of all kinds of reactions are modeled with the Arrhenius expression (Chapter VII). The pyrolysis/mass spectrometry system at MCMS is equipped with a pulsed source that has a time resolution of 100 nanoseconds (0.0000001 second). It can present a series of complete mass spectra as the sample heats up. It is possible to observe changes in the relative rates of complex reactions. Figure VIII-5 shows how products change with time and temperature while running pure image sample 1EB. This kind of presentation is called a "map." The relative rates together with the products' identities tell us a lot about the different components. For example, starch and cellulose are both made of the same glucose units, but starch is much less stable.

The axis numbered 0 - 50 shows the number of an analytical scan. The mass spectrum obtained by each scan is shown on the 50 - 500 axis, the numbers indicating the mass of each fragment. The height of each signal indicates the relative concentration of that ion. Notice that there are no major peaks at masses less than 100 between 30 and 40.

The small graph below shows the total ion current at each time/temperature. The largest amount of products is evolved at the highest temperatures, as would be expected. The map shows some contamination with the xylene that was used to wash adhesive off of the fibers.

55

The same kind of map for the Raes sample is shown in figure VIII-6. The primary product early in the pyrolysis (e.g., between 35 and 40) is m/e 96, furfural. No xylene was used to clean the fibers, because they were not obtained as part of the tape sampling. Product evolution was much lower before scan 30. It is too bad that McCrone did not give us the chance to get such clean spectra on all of the samples.

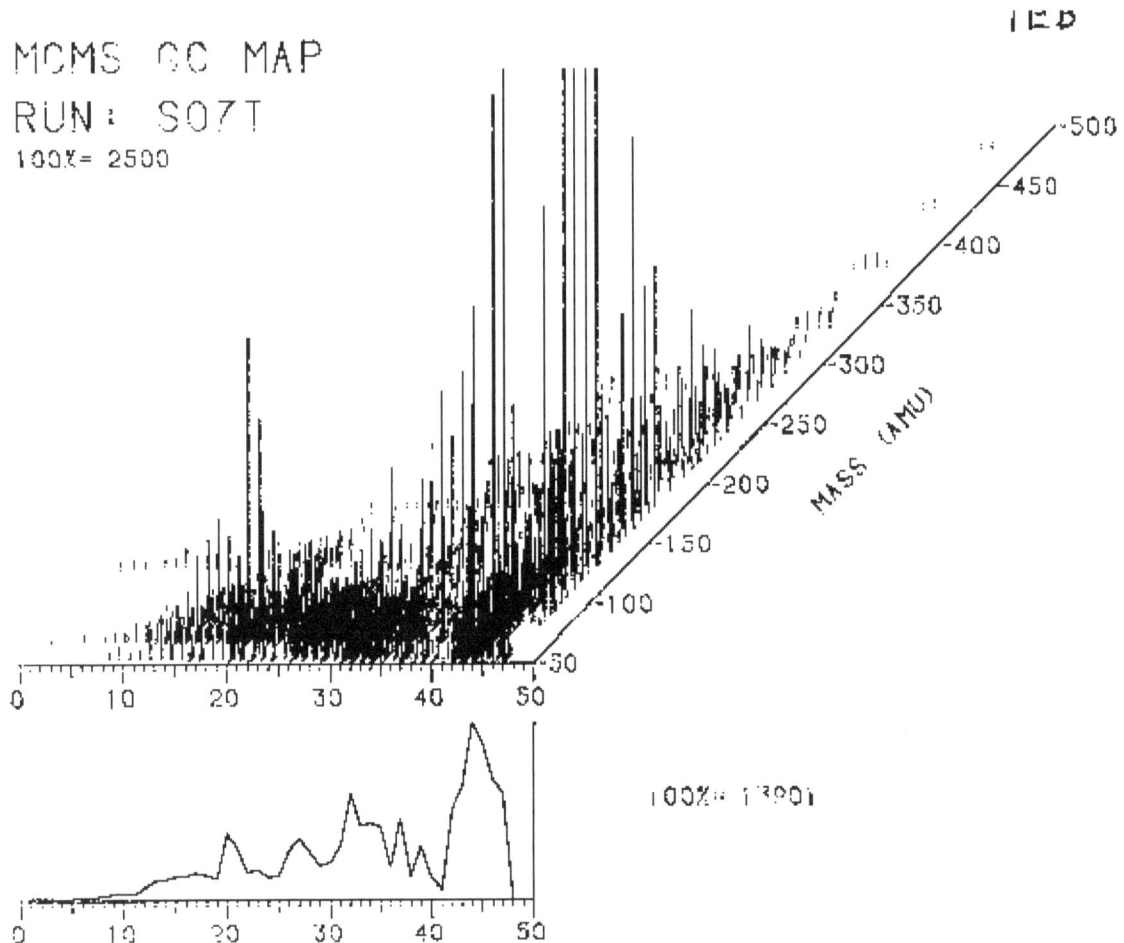

Figure VIII-5: A mass/scan/intensity map of the pyrolysis products from tape sample 1EB, the image of the back of the ankle. Most of the products appear at high temperatures.

Maps of all of the other samples were also obtained. They all showed the same difference in product ratios: the Raes sample was unique. It was contaminated with some material that produced pentose pyrolysis products at relatively low temperatures.

When animal proteins are heated, a major pyrolysis product is 4-hydroxyproline. Hydroxyproline's signal at m/e 131 appeared very early in the pyrolysis of the Zina samples that showed distinct red spots on the fibers (MCMS samples S14S and S15S). Indeed m/e 131 was the major organic pyrolysis product detected between masses 100 and 150 at the lowest temperature that gave any detectable products (figure VIII-7).

56

Mass 131 peaks appeared at much higher temperatures in all of the spectra, but those are in the ranges where cellulose, lignin, and hemicelluloses are decomposing. The spectrum gets very complex at those higher temperatures.

None of the image fibers, control fibers, or Raes fibers showed mass 131 at low temperatures.

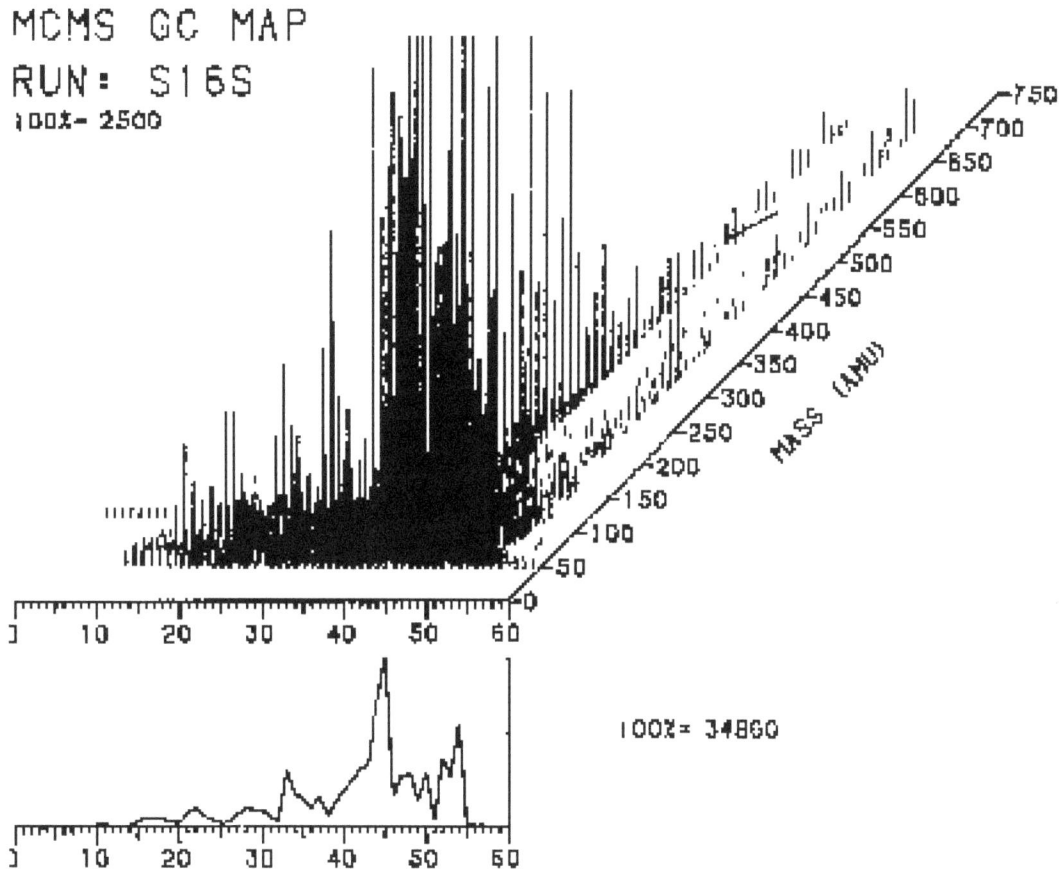

Figure VIII-6: Scan/mass/intensity map for the pyrolysis of the Raes sample. The m/e 96 peak for furfural appears relatively early, and it disappears quickly. The main products come from the pyrolysis of cellulose.

Although mass 131 products can appear from materials other than animal proteins, if animal proteins are present, mass 131 *must* appear in the pyrolysis products. The lower the temperature at which mass 131 appears, the more likely it is a result of the pyrolysis of an animal protein.

Proteins are much less stable than most other natural products. The appearance of a low-temperature emission of hydroxyproline sets a definite upper limit on the highest temperature that could have been seen by the blood after it appeared on the cloth.

The blood was never heated to a high temperature by boiling the Shroud in oil (as sometimes has been suggested) or by the image-formation process.

Unfortunately, we could not detect any products that would explain the colored compound(s) that our eyes saw as image.

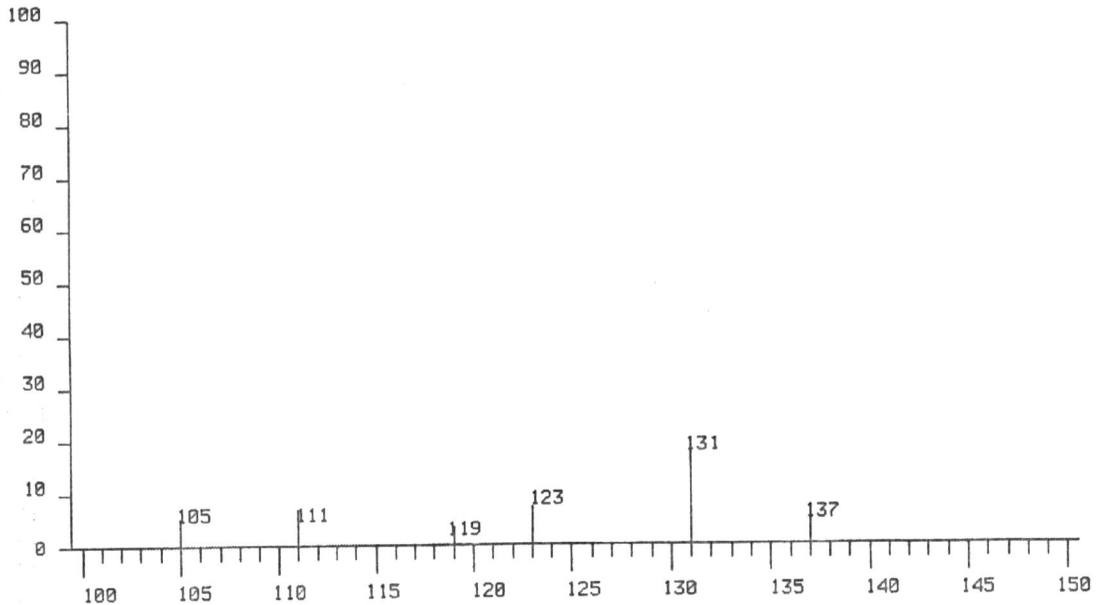

Figure VIII-7: The lowest-temperature mass spectrum that gave detectable products with a blood-spotted Zina sample (S14S).

X-ray Fluorescence Spectrometry:

Morris, Schwalbe, and London assembled the equipment necessary to run x-ray fluorescence analyses on the cloth, and they took the complex system to Turin. This was the most nerve-wracking experiment done in Turin. The equipment failed soon after assembly, and the whole effort would have failed but for the innovative and dedicated help of Franco Faia of Turin, Pasquale Casoli of Milan, and Giovanni Magistrali of Fiat in Turin. I especially enjoyed Franco Faia, a real gentleman, an excellent organist who played the Palace organ for us, and a man whose goal was to play the Radio City Music Hall organ. This was a difficult, time-consuming effort, but it was worth the pain and strain.

X-ray fluorescence spectrometry (XRF) is much like visible-light fluorescence, but the fluorescence comes from elements in the sample rather than organic molecules. Visible fluorescence is emitted when chemical molecules drop from metastable states to a state of lower energy. In XRF, the sample is irradiated with energetic x rays that promote electrons in the inner orbits of elements into higher energy states. When the electrons fall back to lower states, the elements fluoresce at characteristic frequencies in the x ray spectrum. The system used in Turin was capable of observing elements heavier than atomic number 16, sulfur. It could detect and measure all of the expected pigments that would have been used in medieval times or earlier.

Your eye or a camera can not see an amount of pigment that is below a certain detection limit. Above another limit, the surface becomes saturated. The observation of shading requires differences in amounts of pigment. An image such as the Shroud would require a range of concentrations to give the shading observed. Ron London made a series of hematite stains of different densities below 60 micrograms per square centimeter. Schwalbe and London made a calibration curve that compared amounts visible with amounts detected by the XRF. They found that about 2 micrograms per square centimeter of hematite could be seen by the eye.

Total measured iron-compound densities on the Shroud were between about 10 and 58 micrograms per square centimeter. The highest values were in blood areas. They measured the iron concentration of whole blood, and the amount in blood stains was consistent with the measurements.

There was no significant difference in the concentration of any iron compounds from the densest part of the face image into the background. When the face image is compared with the calibration curve, it can be seen that the concentrations of hematite necessary to produce the shading observed with your eye would have been detected by the XRF measurements. This was confirmed by results obtained from x-ray radiographic measurements.

Mottern, London, and Morris reported that they could detect about 7 micrograms of hematite per square centimeter on their x-ray calibration radiographs. No trace of the image could be observed on the radiographs, setting an upper limit on the amount of pigment that could exist on the Shroud. All of the different color densities your eye sees in the image would have to be below that limit.

The x-ray fluorescence measurements, x-ray radiography, visible spectra, and chemical tests are mutually confirmatory. A definite statement can be made that the image was not painted with hematite or any iron pigment.

Morris, Schwalbe, and London did not wish to claim a proof that hematite was not used in any way to enhance the image. They hoped for future XRF tests with isotopic sources instead of an x-ray tube to refine the observations.

On 8 February 1979, Larry Schwalbe reported that he had compared spectra from control areas with blood areas. He was curious about the bromine peak. This got me excited, because bromine is found in Tyrian purple, one of the dyes we thought could have been used to paint or retouch blood spots. He punctured that balloon by adding that the broad peak was the same in control areas. We never found any evidence for significant amounts of any dyes or pigments.

Infrared Spectrometry and Thermography:

Infrared spectrophotometry (IR) is used in chemistry to observe the "functional groups" of different kinds of compounds. Infrared frequencies are those emitted by hot objects, and light in the infrared range interacts with materials in such a way to cause different chemical groups to vibrate, rotate, and wag back and forth. When a wide spectral range of infrared radiation interacts with a functional group, part of the incident energy is absorbed. Infrared spectra show peaks that are characteristic of specific chemical groups, e.g., -OH, C=O, C-Cl, C-H, etc. Unfortunately, solids have very broad peaks, and it is hard to interpret them. Even worse, all carbohydrates have large numbers of -OH and C-H groups, and they tend to absorb huge amounts of the incident energy. We had not expected to get much information from IR reflectance spectra.

The experimental difficulties associated with developing a portable instrument in 1978 and making observations in Turin made the situation nearly impossible. With a detailed knowledge of the problem, Joe Accetta and Stephen Baumgart assembled a single-beam instrument that could operate between about 3-5 and 8-14 microns in the IR. These frequencies miss the major absorptions of the carbonyl group (C=O), one of the most important for analyzing the composition of the image. A single-beam instrument is sensitive to variations in atmospheric composition, and Turin is an industrial city. The computer Accetta and Baumgart took to Turin soon failed; and they were unable to do real-time signal averaging. Only strong minds can remain sane under the conditions they faced.

Their conclusions were the following: "Due to the uncertainties in the data it is not possible to draw definitive conclusions. The spectral similarity of the image areas to known scorches is noted and is consistent with observations in terms of color in the visible region of the spectrum; however, this result is not without ambiguity since spectral similarities are characteristic of most areas examined as shown by the data in both spectral bands... Without drastic improvements in instrumentation, a second attempt is not recommended."

They remarked that surface effects overwhelmed chemical differences. My personal opinion is that drastically improved infrared equipment would not yield useful information from the solid surface of the Shroud.

Accetta and Baumgart also reported results from thermography measurements. Thermography allows the detection and imaging of small temperature differences on a surface. Different materials emit different amounts of radiant energy when they are at the same temperature. Differences in the composition of the image on the Shroud might have appeared in differences in emissivity. One of their conclusions was the following: "We have shown that emissivity differences in various features of the Shroud are too small to yield recognizable images in the 3-5 or 8-14 micrometer band with instruments of temperature sensitivity on the order of ... 0.5°C or less." When the surface of the Shroud was illuminated with flood lamps, the temperature span from black to white areas was about 1.75°C. Under those conditions they got good resolution of the image in the 8-14 micrometer wavelength region. All of the image seemed to be the same. Their final

60

conclusion was the following: "With due regard to the limits of instrument resolution and sensitivity, it is the authors' opinion that no significant anomalies exist."

My interpretation of their results is that, whatever produced the image produced identical surfaces in both the front and back images. It is also obvious that paints were not mixed to produce shading.

Laser Microprobe Raman Spectrometry (MOLE):

Raman spectrometry is much like IR spectrometry; however, it sees different motions of the chemical functional groups. It provides a good complement to IR. Unfortunately, we could not make the observations in Turin, but fibers from the tape samples could be analyzed in the US.

Joan Rogers identified suitable fibers on the tape samples and prepared them for analysis. She took tapes, fibers from non-image areas, and fibers from image areas to Instruments SA, Inc., in Metuchen, N.J. in December 1979. The samples were analyzed by Dr. Fran Adar. Similar samples were analyzed by Mark Anderson, McCrone's MOLE expert in January 1980.

Anderson observed that most of the red flecks on the Shroud "bubbled up and turned black" when he hit them with the laser beam. This was an entirely different response than he got from authentic hematite crystals. He said it "acted like an organic phase" (21 January 1980). Walter McCrone refused to accept those observations. If he wanted the image to be painted with hematite, no conflicting observations would be allowed.

It was easy for the microprobes to detect the Mylar backing on the sampling tapes, but no quantitatively significant Raman spectra could be obtained from any of the samples. There was no evidence for any chemical products from *Saponaria officinalis* or any other coating on image fibers.

UV and visible spectrometry would not see significant differences among the carbohydrates. The -OH vibrational states of all of the carbohydrates and water are very broad and intense, and neither IR nor Raman spectrometry could distinguish among them. We were not looking for trace carbohydrate impurities, we were looking for painting-type impurities on the cloth.

All of the observational methods agreed that no pigments, normal painting vehicles, or natural exudations (other than the blood) had been added to the cloth after its production. The image on the Shroud of Turin is not a painting. No foreign materials were added to the cloth in image areas.

CHAPTER IX: THE RADIOCARBON SAMPLE

The Shroud has been controversial since it was first documented in the 14th Century. If we knew exactly how old it is, much of the controversy would be eliminated. But nothing is ever simple with the Shroud.

It has been possible to determine the age of natural organic materials by radiocarbon analysis since the 1950s. During the first meeting to discuss scientific studies on the Shroud of Turin, I pushed for radiocarbon dating [R. N. Rogers, "Chemical Considerations Concerning The Shroud of Turin," in Kenneth Stevenson (Ed.), *1977 United States Conference of Research on The Shroud of Turin,* 23-24 March 1977, Albuquerque, NM, USA, Holy Shroud Guild, 294 East 150 St., Bronx, N. Y. 10451 (pp. 131-135).] There was great antagonism to the suggestion, and it is still a very sore point with the custodians of the Shroud. They have a good reason to be antagonistic: the first attempt to determine a radiocarbon age was very poorly managed in Turin, and it proved to be a major fiasco.

Carbon-14 is a radioactive isotope of carbon (radiocarbon). It is continually produced in the upper atmosphere by the collisions of energetic neutrons, produced by cosmic rays, with nitrogen-14. The carbon-14 atoms rapidly become part of carbon dioxide molecules. The carbon dioxide is taken up by green plants in the process of photosynthesis. As long as the plant is alive, its radiocarbon level is constant. When the plant dies, the radiocarbon is not replenished, and it decays at a known rate by emitting a beta particle (an energetic electron). Half of the radiocarbon decays every 5,770 years; therefore, a measurement of the radiocarbon left in a sample of plant material tells how long it has been since the plant died. This method of age determination (usually called "dating") was developed by Willard Libby, and it provided the ideal method for determining how old an ancient piece of linen was.

Originally, radiocarbon was measured by counting its disintegrations in a radiation detector. The process required several grams of carbon, and it would have been destructive to the Shroud. STURP considered a radiocarbon age determination to be the most important analysis that could be run on the Shroud; however, the custodians would not consider even the most sensitive methods available in 1977.

More recently, methods have been developed that use a nuclear accelerator to provide the analog of a mass spectrometer. The isotopes of carbon can be separated and measured independently. Very small samples can be analyzed. In 1988, the custodians allowed samples to be cut for radiocarbon age determination, and they were run at three of the best laboratories in the world; the University of Arizona, the University of Oxford, and the Institut für Mittelenergiephysik, Zürich.

The 1988 radiocarbon analyses were the best that could have been obtained anywhere in the world. Effects of sample-preparation methods (cleaning) were studied. The three laboratories analyzed sample aliquots, and they reported that "The age of the shroud is obtained as AD 1260-1390, with at least 95% confidence." Unfortunately, that date does not reflect the STURP observations on the linen-production technology and the chemistry of the fibers from the tape samples.

In many cases where questions arise, an appeal is made to "authority." There can be no question about the authority of the radiocarbon investigators; however, true scientists like to see all loose ends questioned and tested. With the Shroud, neither the radiocarbon investigators nor the authorities in Turin have cooperated in attempts to resolve the "dating problem." The church officials appear to be content to have society view the Shroud as a medieval hoax, and the radiocarbon laboratories have refused to consider the possibility that they were given a spurious sample. In a manner uncharacteristic of rigorous scientists, they refuse to allow observations on retained samples. They also refuse to do their own simple chemical observations. They refuse to discuss or show any photomicrographs of samples they might have. This kind of action is all too characteristic of Shroud studies. Emotions tend to overwhelm science.

It is common in archaeology for a radiocarbon date to be a surprise. The surprise is usually a result of a "lack of association." The radiocarbon sample does not represent the date of the event in question. In one case, the radiocarbon date obtained from charcoal in a fire pit was thousands of years older than the artifacts littering the floor of the cave. The archaeologist noticed that the people who made the artifacts had been burning wood from a packrat midden in the back of the cave. The midden was partially fossilized, and it dated from the last ice age. There was no association.

The 1988 sampling operation was described as follows: "The shroud was separated from the backing cloth along its bottom left-hand edge and a strip (~10 mm x 70 mm) was cut from just above the place where a sample was previously removed in 1973 for examination (the Raes Sample). The strip came from a single site on the main body of the shroud away from any patches or charred areas." The location of the sample is shown in figure IX-1.

Unfortunately, the sample was approved at the time of sampling by two textile experts, Franco Testore, professor of Textile Technology at the Turin Polytechnic, and Gabriel Vial, curator of the Ancient Textile Museum, Lyon, France. No chemical or microscopic investigations were made to characterize the sample. I believe that was a major disaster in the history of Shroud studies. Control samples should always be retained to enable confirmation of results at a later date. Retained samples, if any, have not been made available for study. This leads one to question the ethics or rigor of any "scientists" involved in the process. Is something being hidden?

I had archived samples from the sampling tapes, the Raes sample, and the Holland cloth and patches after STURP disbanded. These samples were available for testing the validity of the radiocarbon sample. Fortunately, the Holland cloth provides an authentic, documented sample of medieval linen. It should provide an example of the type of linen available at the time suggested by the radiocarbon date. These samples could provide a rather convincing argument for the properties of the radiocarbon sample; however, a definitive statement could not be made until I received yarn segments from the authentic radiocarbon sample on 12 December 2003. Now we could talk dating with ample proof.

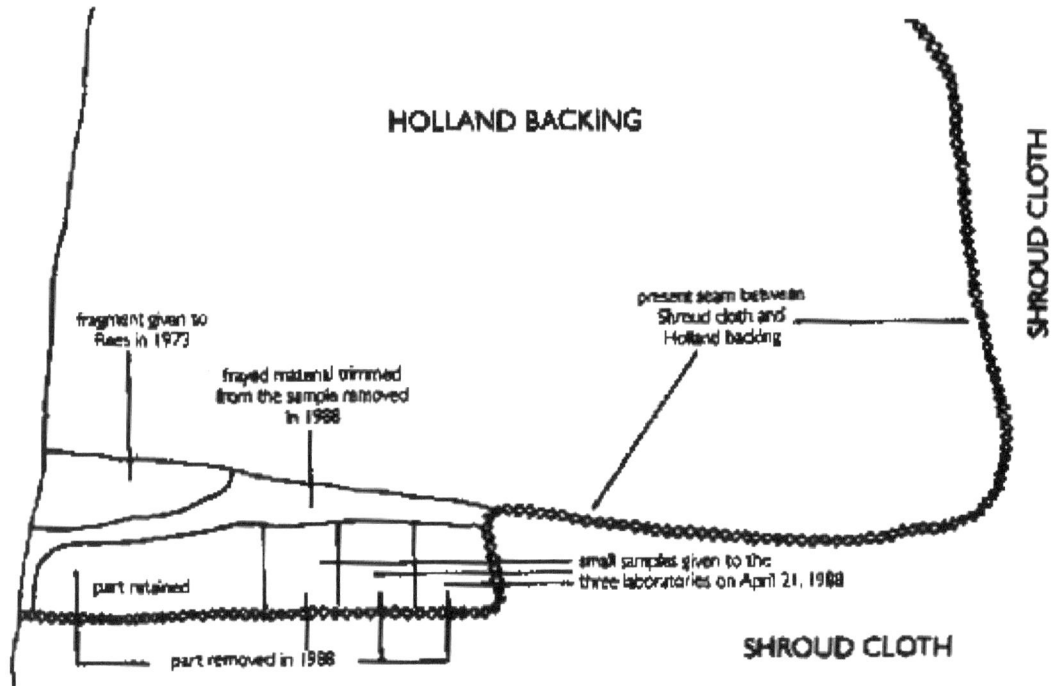

Figure IX-1: Locations of Raes and radiocarbon samples (bottom margin of cloth is to left). There should be compositional similarities between them. Retained samples should be studied by chemistry and microscopy.

Samples:

Professor Gilbert Raes of the Ghent Institute of Textile Technology cut a small sample from the cloth in 1973[4]. The location of his sample is shown in figure IX-1. His sample contained yarn segments from both sides of the seam that can be seen running vertically from his sampling area (figure IX-11). He called the sample on the right of the seam Part I, and that on the left Part II. He found that Part I contained cotton, and he reported that the cotton was an ancient Near Eastern variety, *Gossypium herbaceum*, on the basis of the distance between reversals in the tape-shaped fibers (about eight per centimeter). He did not find any cotton in Part II.

I received 14 yarn segments from the Raes sample from Professor Luigi Gonella (Department of Physics, Turin Polytechnic University) on 14 October 1979. I now have these samples numbered, photographed, and identified as the "Raes threads." There was no indication which segments came from Part I and which, if any, from Part II; however, all are longer than would be expected for segments from Part II.

On 12 December 2003, I received samples of both warp and weft threads that had been taken from the radiocarbon sample by Professor Luigi Gonella before it was distributed for dating. He reported that he excised the threads from the center of the radiocarbon sample. A "chain of evidence" has been maintained on those threads, and it is certain that they were truly removed from the radiocarbon sample. These samples finally made it possible to confirm my conclusion that the radiocarbon sample was not valid.

As described by Damon et al., the radiocarbon sample was cut from directly above the Raes sample: The two areas should have similar chemical compositions; therefore, I made a concerted study on the larger Raes sample, saving the authentic radiocarbon sample for confirmation. I wanted to make certain that sufficient sample remained for independent confirmation of my observations. Nothing involved with the Shroud should ever be accepted without independent confirmation.

As part of The Shroud of Turin Research Project (STURP), I took adhesive-tape samples from all areas of the Shroud in 1978[2]. The tape was produced specifically for the project by Ronald Youngquist of the Minnesota Mining and Manufacturing Corporation. He used an amorphous, pure-hydrocarbon adhesive that would not contaminate the Shroud or the samples, and the adhesive could be removed by washing with xylene. The tapes were applied to the surface of the Shroud with a pressure-measuring applicator to enable semi-quantitative comparisons among samples.

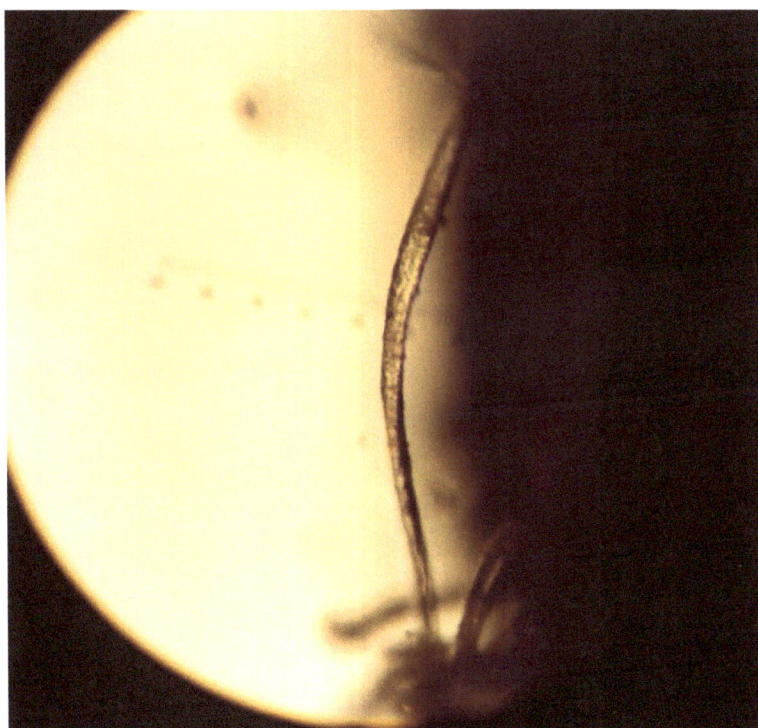

Figure IX-2: Heavily encrusted cotton fiber emerging from Raes #14 (400X).

The Shroud was badly damaged in a church fire in AD 1532. Nuns patched burn holes and stitched the Shroud to a reinforcing cloth that is now known as the Holland cloth. I also sampled it in 1978. The Holland cloth provides an authentic, documented sample of medieval linen.

I archived samples from the sampling tapes, the Raes sample, and the Holland cloth and patches after STURP disbanded. These samples were available for testing the validity of the radiocarbon sample. Fortunately, the Holland cloth provides an authentic, documented sample of medieval linen. It should provide an example of the type of linen available at the time suggested by the radiocarbon date.

Observations:
As reported by the radiocarbon investigators and shown in the diagram of the sampling area (figure IX-1), the Raes sample was cut from immediately below the radiocarbon sample. It should share at least some warp yarns with the dating sample. A much larger sample of Raes threads is available for study than can be obtained from the radiocarbon laboratories or the custodians of the Shroud. However, observations on the Raes sample should give an indication of the properties of the radiocarbon sample. The Raes threads can provide most of the information we need to test the validity of the radiocarbon sample. If the properties of the Raes sample and the main part of the Shroud are different, the radiocarbon sample should be invalid. Thanks to the December 2003 sample, we can now confirm all observations with a documented radiocarbon sample.

(A) Cotton is not common in Shroud samples.
Cotton fibers are easy to find mixed intimately with the linen fibers of Part I Raes threads. Figure IX-2 shows a heavily encrusted cotton fiber on the surface of one of the Raes threads. It can be identified by its flat, tape-like shape, the presence of one reversal, and the absence of the bamboo-like growth nodes of linen. When the cotton fiber was drawn out of the thread, it showed reversals about 1.2-mm apart. This agrees with Raes' observations.

Figure IX-3 shows fibers from the radiocarbon sample. The flat ones with a twist in them are cotton. Notice that both cotton fibers are completely covered by a colored layer. Some of the linen fibers are nearly clean. Also notice that the linen fibers have very little lignin at their growth nodes. Indeed, the growth nodes are so clean you need polarization to see them (figure IX-4). The radiocarbon sample contains cotton, the fibers are coated, and the bleaching method was more efficient than that used on the main part of the Shroud.

I did not attempt to make a quantitative cotton comparison between Raes and radiocarbon threads and Shroud tapes, because there was too little cotton of any kind on Shroud samples. We had been puzzled by the Raes report at the time of the 1978 STURP observations in Turin. We could not find more than traces of cotton on the Shroud. The cloth appeared to be pure linen.

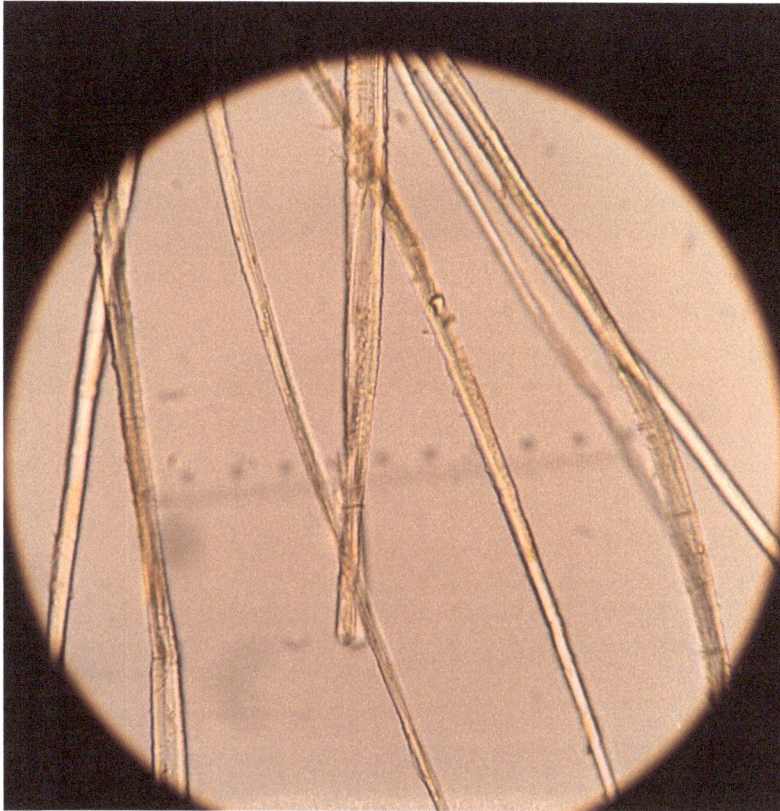

Figure IX-3: Cotton and linen fibers from a warp thread of the radiocarbon sample, 800X in 1.345-index oil.

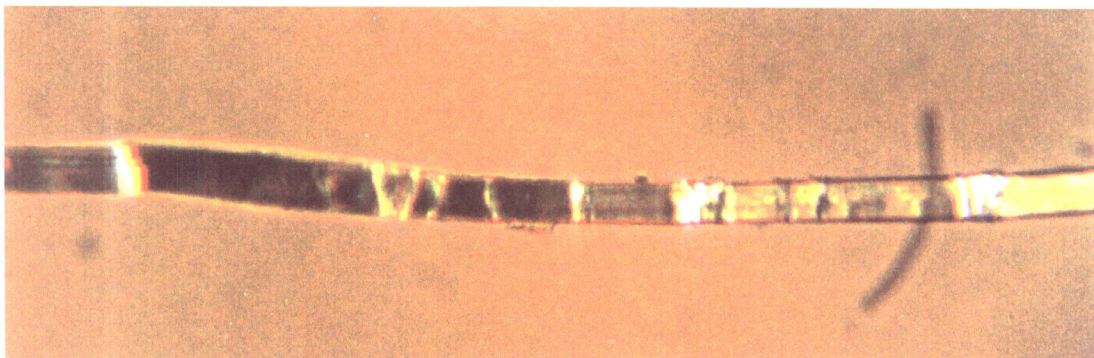

Figure IX-4: A radiocarbon-sample warp fiber between crossed polarizers, 800X in 1.345 oil. The growth nodes rotate polarized light differently than does the body of the fiber. The birefringence color depends on the angle of the fiber *versus* the angle of the polarized light.

Fibers retained on the sampling tapes can be differentiated according to their indexes of refraction compared with the index of the tape's adhesive. The two indexes of cotton are close to that of the adhesive. Birefringence is first-order white. The index of linen across the fiber is appreciably lower than that of the adhesive. It is more birefringent than the cotton.

We used cotton gloves during the STURP studies of 1978 to protect the relic, and they could have been responsible for the traces of modern cotton found on a few Shroud sampling tapes.

Samples from the main part of the cloth are significantly different from the Raes samples and the radiocarbon sample with regard to cotton content.

(B) Amounts of lignin differ between Raes/radiocarbon samples and Shroud fibers.

The linen fibers found on Shroud tapes average about 13-μm diameter. They show periodic growth nodes, and they look like microscopic lengths of bamboo. Figure V-3 shows several linen fibers that were pulled from the image at the back of the ankle. It is a completely unpolarized photograph. There is no dichroism or birefringence color. These fibers are characteristic and representative of image fibers. There are dark deposits of lignin on most of the growth nodes. Absolutely no cotton could be found among the hundreds of fibers on this tape sample.

As figure IX-5 shows, very little lignin is visible at the linen growth nodes of the Raes and Holland cloth samples. Lignin is a dark, complex natural structural polymer that is found in all woody plants. Its composition and structure are specific to a given plant, but phenolic units are common to all lignins. It is not a polysaccharide (polymer composed of sugar units) like starch and cellulose. Linen is bleached to remove lignin; however, it is unusual to find a Shroud fiber without some significant deposits of lignin.

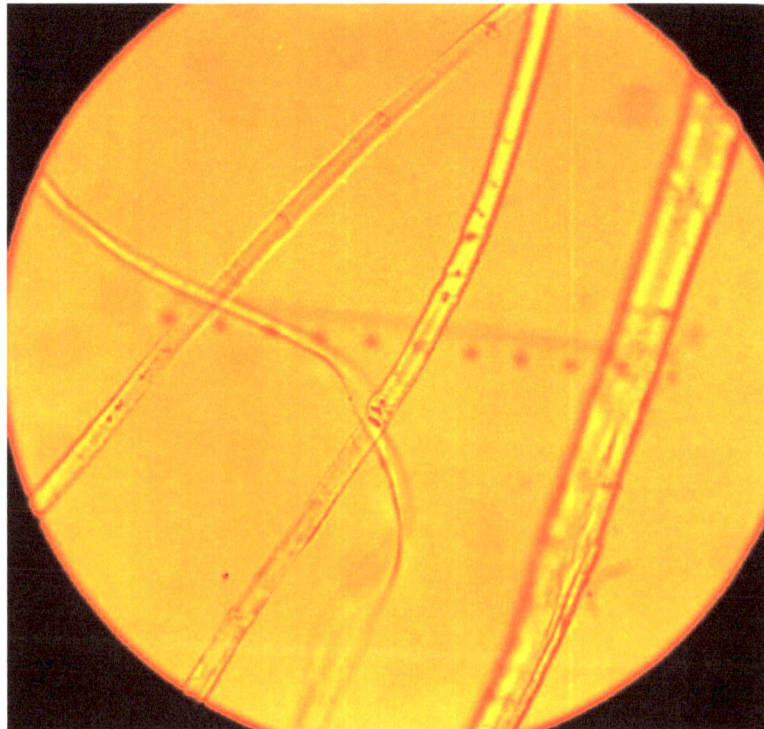

Figure IX-5: Fibers from Raes #5 mounted in 1.515 oil. Very little lignin is visible at the growth nodes.

(C) Quantitative evaluation of lignin.

Simple microscopic viewing is not sufficient to prove differences among the samples. In order to obtain quantitative data, I counted hundreds of growth nodes in each sample and noted which showed traces of lignin.

The table shows that fibers from the Raes threads, Holland cloth, and modern linen show very little lignin at growth nodes, and the amounts of lignin in those samples are quite consistent. Notice that the numbers refer to *percentages* not numbers of growth nodes observed. No samples of the Holland cloth or Raes threads had heavy deposits of lignin.

Unlike the Raes and Holland cloth samples, the fibers on the Shroud tapes vary greatly in amounts of lignin. A large number of observations shows that lignin ranges from heavy to nil, depending on the location from which the sample was taken. There is an explanation for this observation.

SAMPLE	NODES WITH LIGNIN (%)
Modern Commercial	55, very light
Raes Threads	40, light
Holland cloth	60, light
Right Foot, Dorsal Image	54, heavy to moderate
Finger, Frontal Image	80, light
Ankle, Dorsal Image	100, heavy to moderate
Scorch control	39, heavy to moderate

(D) Lignin amounts vary among Shroud locations.

The bands of color seen in Shroud photographs have been mentioned. Both warp and weft yarns show this property. Some areas show darker warp yarns and some show darker weft yarns. In some places, bands of darker color cross. In other places, bands of lighter color cross. The effect is somewhat like a plaid.

All of the bleaching processes used through history remove lignin. The more quantitative the bleaching process the whiter the product. The bands of different color on the Shroud are the result of different amounts of lignin left from the bleaching process. The tape samples reflect this variation as observed differences among quantitative measurements of lignin on the fibers.

Anna Maria Donadoni pointed out locations where batches of yarn ended in the weave and new yarn had been inserted in order to continue weaving. The yarn ends were laid side by side, and the weave was compressed with the comb. The ends are often visible, and the overlaps appear to correspond to zones of different color in the weave. The different batches of yarn show different colors.

Differences among amounts of lignin on linen fibers in the Raes and radiocarbon samples and on Shroud fibers are significant. There is a similar difference among other impurities on the Raes and radiocarbon samples and the main part of the Shroud.

(E) Raes and radiocarbon threads show a yellow-brown coating.

Raes and radiocarbon threads show colored encrustations on their surfaces. Some sections of medulla contain some of the material, showing that it had been able to flow by capillary attraction as a liquid. The encrustation is not removed by nonpolar solvents, but it swells and dissolves in water. There was absolutely no encrustation on either the Holland cloth or fibers from the main part of the Shroud.

Figure IX-4 shows two cotton fibers from Raes thread #5. One of the fibers was taken from inside the thread, and it is nearly colorless. The other fiber was taken from just under the outer surface of the thread, it is deeply colored, and it shows gelatinous material adhering to its surface. A marked difference between inside and outside fibers is characteristic of Raes samples.

The outside of Raes thread #14 showed the heaviest encrustation and deepest color of any of the samples. The encrustation is heaviest on cotton fibers, it is the vehicle for the yellow-brown color, and it suggests that the cotton was added to enable better control of dyeing or staining operations.

When I teased threads open at both ends with a dissecting needle, the cores appeared to be nearly colorless. This observation suggests that the color and its vehicle were added by wiping a viscous liquid on the outside of the yarn in order to match the color of new material to the old, sepia color of the Shroud.

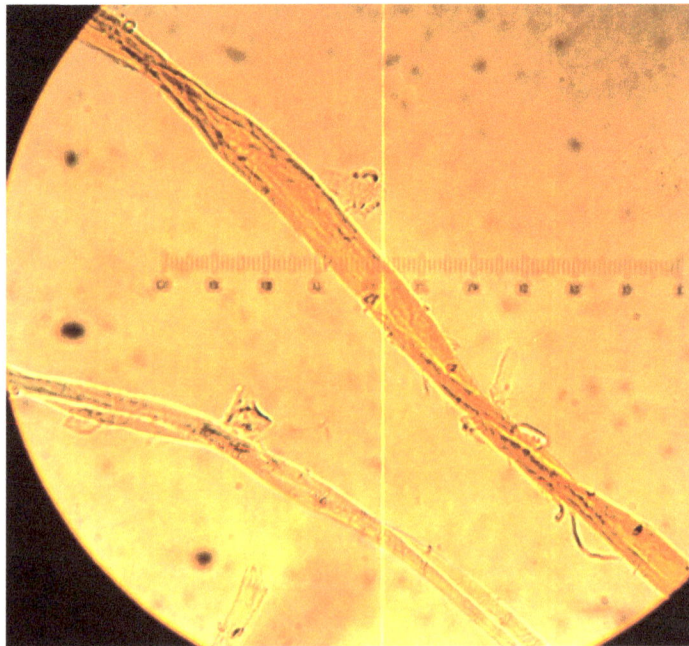

Figure IX-4: Two cotton fibers (X400) from a Raes thread, one nearly colorless from inside and one encrusted and red from outside.

70

The yellow-brown encrustation shown in figure IX-5 swelled and became more transparent as it soaked. The color instantly changed to bright yellow in 6N hydrochloric acid (HCl), and the coating was reduced in density as the fibers were soaked in the acid (figure IX-6).

The natural dye extracted from Madder roots was important in the Near East for thousands of years. It appeared in Italy about the time of the last Crusade, but it was not until the 16th Century that it appeared in France and England. The first European book on dyeing was published in AD 1429.

Spots of colored dye on a mordant are called "lakes." Bright red lakes of dye were found on many of the Raes fibers, indicating that at least some Madder root dye was used and that some of the color appeared on a hydrous-aluminum-oxide mordant. Some purpurin appears in Madder root extract, and it reacts much the same as alizarin. Hydrous aluminum oxide is instantly soluble in 6N HCl, and alizarin is bright yellow in acid (figureIX-6).

The presence of aluminum in the coating material is consistent with the results of Adler, Selzer, and DeBlase. These researchers performed elemental analyses on different shroud materials, including fibers from warp threads taken from the radiocarbon sample. They reported concentrations of aluminum on the radiocarbon sample 20-times those on shroud fibers. The observation of alizarin/hydrous-aluminum-oxide lakes on Raes and radiocarbon samples together with the elemental analysis proves that the samples had been dyed.

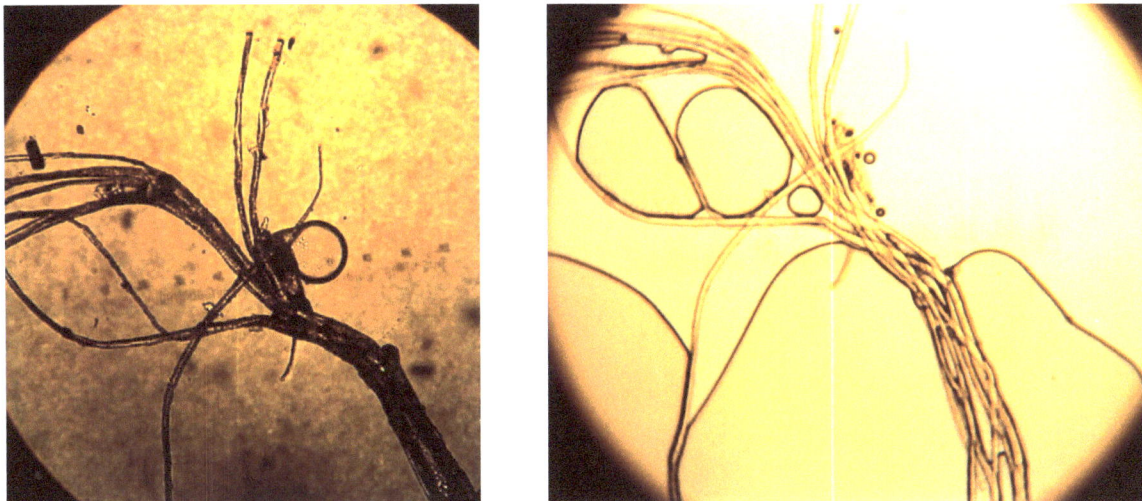

Figure IX-5: (Left) Heavily encrusted fibers from the outside of Raes #14 (400X) mounted in water.

Figure IX-6: (Right) The same fibers shown in figure IX-5 mounted in 6N HCl. Notice the bright yellow color.

Alizarin is used as an acid-base (pH) indicator in chemical analysis. It is yellow below a pH of 5.6 and red above a pH of 7.2 (figure IX-7), changing to purple above 11.0 (figure IX-8). This agrees with observations on the coating. Madder root dye is a highly probable contributor to the color of the coating on the Raes samples. No dye could be detected on any image fibers.

As mentioned in Chapter VII, I had obtained a positive test for pentose sugars on fibers from the Raes sample with Bial's reagent (orcinol, con. HCl and $FeCl_3$). I could not get a positive test from Shroud samples. The fact that the gum hydrolyzed in Bial's reagent (made with con. HCl) to give a pentose test should have given us a clue in 1980 that the Raes sample was different from the main part of the Shroud. Sometimes understanding comes slowly.

However, not all of the polysaccharides on the Raes fibers were removed by concentrated HCl. Higher-molecular-weight starch fractions are much more difficult to hydrolyze than are polypentose-containing plant gums. Some starch could be detected on HCl-cleaned Raes fibers with an aqueous iodine reagent.

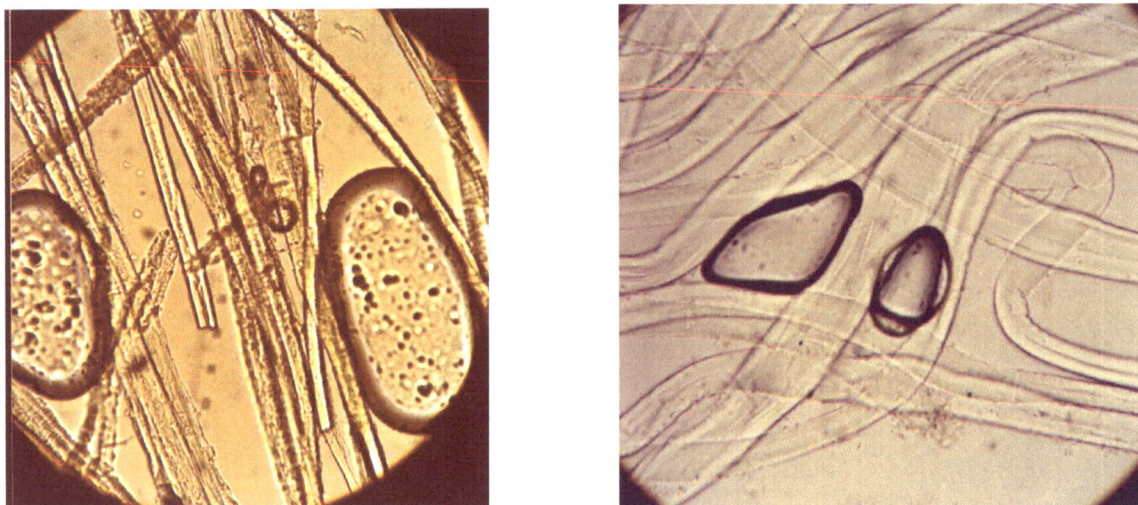

Figure IX-7: (Left) Surface fibers from Raes #14 reddened in $NaHCO_3$ at a pH of 8.0.

Figure IX-8: (Right) Surface fibers from Raes #14 turned purple by soaking in a high-pH medium, 2N NaOH.

I arranged two heavily-encrusted fibers from the outer surface of Raes #5 across each other and covered them with a cover slip. The dry fibers were nearly opaque as a result of the coating. I then ran aqueous iodine solution under the cover slip by capillary flow. The iodine quickly turned the coating bright yellow, indicating a plant gum. The coating swelled and partially dissolved in the water. I let the water and iodine evaporate overnight. The redeposited, colorless, gelatinous material is clearly visible along the fibers in figure IX-9. The iodine was in simple solution in the gum. It did not produce the yellow color by iodination or iodine-catalyzed dehydration reactions.

The horizontal cotton fiber in figure IX-9 shows a deep-red coloration. When tested with iodine, normal soluble starch turns blue. Starch that is soluble only in hot water turns red. The higher-molecular-weight, hot-water-soluble starch is the last to wash out of a cloth.

The encrustation on Raes samples is almost certainly a plant gum. The gum does not appear on any of the other linen samples that are associated with the Shroud of Turin. It is highly probable that the alizarin-dyed, gum-coated yarns extend into the adjoining radiocarbon samples.

L. A. Garza-Valdes was allowed to observe fibers from a part of the radiocarbon sample that had been retained by Giovanni Riggi di Numana after the sampling operation. He reported seeing a coating on the fibers, and a photomicrograph has been published on the web. The photomicrograph shows a yellowish coating on a linen fiber; however, the provenience of the fiber is not specified. He misinterpreted the coating as a "bioplastic polymer." Chapters VII and VIII detail reasons why it could not be a bioplastic polymer. However, his observations increase the probability that the radiocarbon sample has properties identical to those of the adjoining Raes sample.

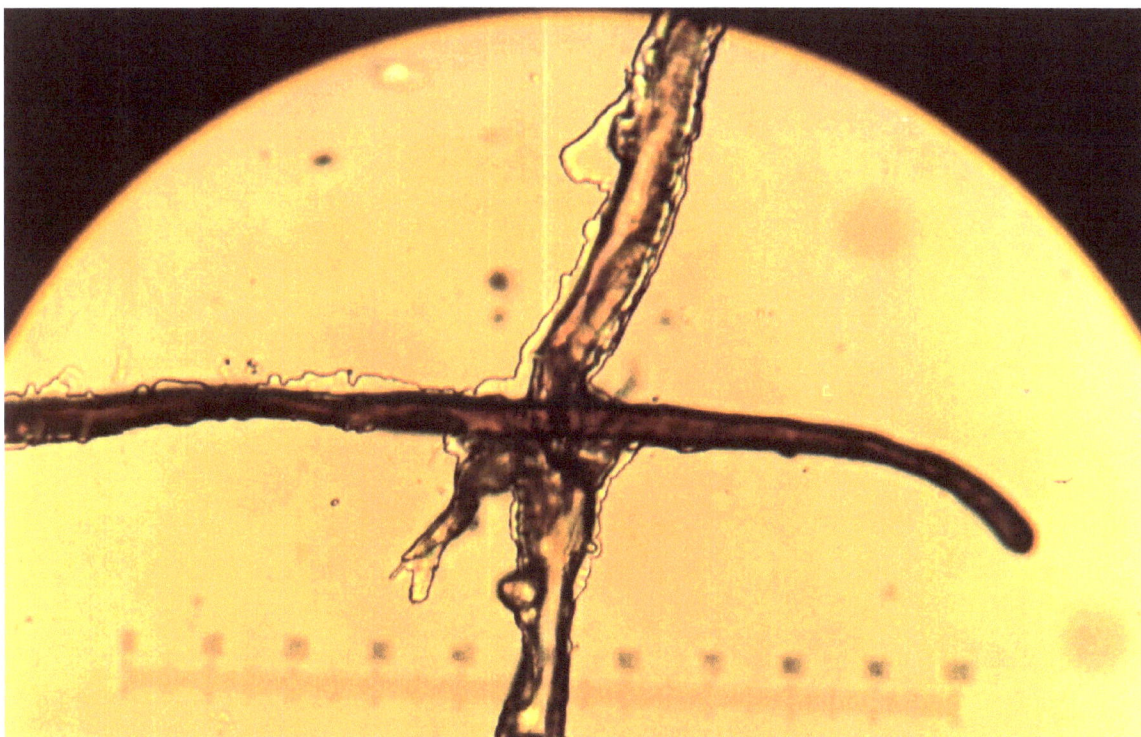

Figure IX-9: Encrusted cotton fibers from the outer surface of Raes sample #5 (400X) after treatment with aqueous iodine and drying.

The gum is probably the same age as the Raes threads, and it should have had no effect on the age determination. In any case, it would also have been removed by the cleaning procedures used on the dating sample. However, the presence of a gum coating on retained 1988 radiocarbon-dating samples would prove that the samples were not representative of the main part of the relic's cloth. Such a lack of association would prove that the radiocarbon date is invalid.

The relatively easy water solubility and hydrolysis of the encrustation suggests gum Arabic. It is obtained from *Acacia senegal*, and it is mostly composed of pentose-sugar units. It turns bright yellow in aqueous iodine, as observed on the Raes threads. Gum Arabic has been used for thousands of years, and it is still used in inks, textile printing, and the adhesive on postage stamps.

(F) The Raes samples show a unique splice.

Raes thread #1 (figure IX-10) shows distinct encrustation and color on one end, but the other end is nearly white. The photograph was taken on a 50% gray card for color comparison. Fibers have popped out of the central part of the thread, and the fibers from the two ends point in opposite directions. This section of yarn is obviously an end-to-end splice of two different batches of yarn. No splices of this type were observed in the main part of the Shroud.

Figure IX-10: Raes thread #1 showing an end-to-end splice. The two ends show different colors and amounts of coating.

(G) Other observational methods show anomalies in the radiocarbon sample area.

The specific area where the radiocarbon sample was obtained was photographed in 1978 with visible light, low-energy x rays at high resolution, transmitted light, and a pure ultraviolet source. All of these photographs were available before the Shroud was sampled for radiocarbon analyses. They must have been totally ignored. I believe that the sampling area was one of the worst that could have been chosen.

While making the UV photographs (figure IX-11), the source was heavily filtered to exclude visible light and the camera was heavily filtered to exclude any effect of the UV on the film. All that appears on the film is the result of pure fluorescence.

The radiocarbon sample was cut from the area that is immediately above the white, triangular-shaped feature (figure IX-11). It is darker than normal, a fact that is not the result of image color or scorching. The cloth is much less fluorescent in that area, brightening into more normal fluorescence to the right. The features to the right are a scorch and a water stain. The ultraviolet-fluorescence photograph proves that the radiocarbon area has a different chemical composition than the main part of the cloth, and it is truly anomalous.

The normal non-image cloth shows weak fluorescence (figure IX-11 upper right). When image appears on the cloth, it appears brown over the background fluorescence of the cloth.

The low-energy, high-resolution x-ray transmission photograph (figure IX-12) was made in 1978, ten years before the radiocarbon sample was cut. It shows areas of higher density as light-gray streaks. The "banded" characteristic of the cloth is easy to see; however, some bands do not extend from the main part of the cloth into the radiocarbon-sample area: There is a different "plaid" pattern in that area.

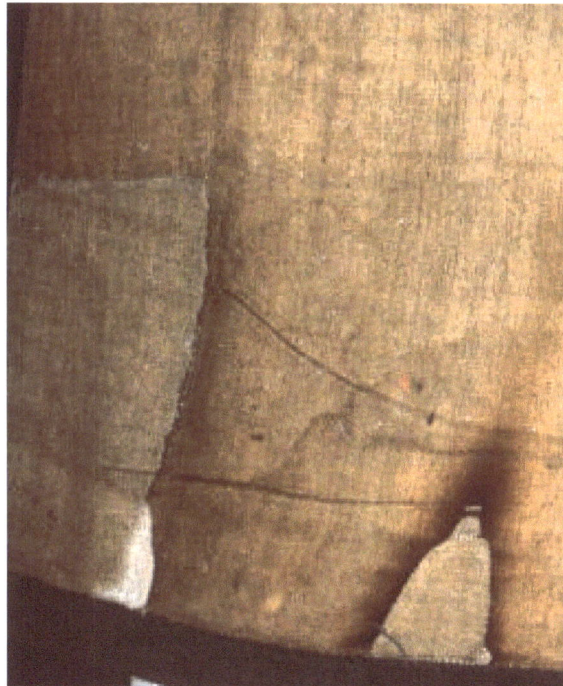

Figure IX-11: UV fluorescence photograph of the radiocarbon-sample area. The small, white area is the location of the Raes sample, which adjoins the radiocarbon sample. ©1978 Vernon Miller

Area 1 is the Raes sample. The radiocarbon sample was cut from between the Raes sample and the upper double crease (figure IX-12). Both creases are double on the x ray, because they appear in both the Shroud and Holland cloth; i.e., they formed after the fire of AD 1532. Area 2 was cut off long ago as a "souvenir," and it shows only the Holland cloth backing (low-density, black with no banding). Lines a, b, and c are continuous bands of different density that extend across the cloth. The radiocarbon area shows anomalous banding.

Conclusion on the association between the radiocarbon date and the time at which the Shroud was produced:

The combined evidence from chemistry, cotton content, technology, photography, ultraviolet fluorescence and residual lignin proves that the material of the main part of the Shroud is significantly different from the radiocarbon sampling area. The validity of the radiocarbon sample must be questioned with regard to dating the production of the main part of the cloth. A rigorous application of Scientific Method would demand a confirmation of the date with a better selection of samples.

The Shroud was subjected to a major "restoration" in June and July of 2002. During the operation, considerable charred material was scraped from around the burn holes. That material would be ideal for radiocarbon dating. The published date can not be accepted without confirmation.

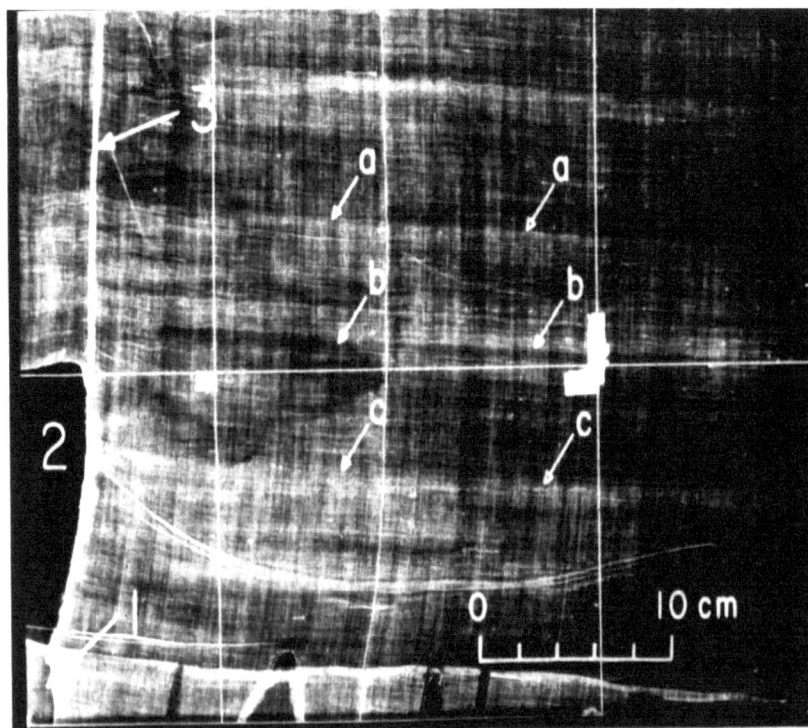

Figure IX-12: Low-energy, high-resolution x-ray photograph of the Radiocarbon - sample area. Areas numbered are the following: 1, the Raes sample; 2, a cutout portion of the Shroud where only the Holland cloth is visible; 3, a seam parallel to the long dimension of the cloth. Alternating high- and low-density bands that continue across the seam are indicated by the arrow sets labeled a, b and c. The radiocarbon sample was cut upward from 1.

CHAPTER X: IMAGE-FORMATION HYPOTHESES

Any serious image-formation hypothesis must explain as many of the recognized and published facts as possible. It is not valid to eliminate or suppress facts in order to support a specific hypothesis. Hypothesis testing is an ongoing process: All new, confirmed observations must be included while testing existing hypotheses and suggesting the development of new hypotheses. A concomitant attempt to recognize pertinent natural laws should be pursued, and all hypotheses should be equally tested against them.

SPECIFIC IMAGE-FORMATION FACTS FOR TESTING HYPOTHESES:

Any hypothesis for image formation must agree with the laws of physics and chemistry and explain all of the different types of controlled and/or quantitative scientific observations. "I-think-I-see" observations are not acceptable. A list of confirmed facts follows.

1) Reflectance spectra, chemical tests, laser-microprobe Raman spectra, pyrolysis mass spectrometry, and x-ray fluorescence all show that the image is not painted with any of the expected, historically-documented pigments.

2) No painting pigments or media scorched in image areas or were rendered water soluble at the time of the AD 1532 fire.

3) Direct microscopy showed that the image color resides *only* on the topmost fibers at the highest parts of the weave.

4) The color density of any specific image area depends on the batch of yarn that was used in its weave. The cloth shows bands of slightly different colors of yarn.

5) Adhesive-tape samples show that the image is a result of concentrations of yellow-brown fibers.

6) The image does not fluoresce under ultraviolet illumination.

7) The image of the dorsal side of the body shows the same color density and distribution as the ventral, and it does not penetrate the cloth any more deeply than the image of the ventral side of the body.

8) Thermography proved that the emittance of the image was the same in all areas. The entire image formed by the same mechanism. Spectra and photography confirmed this observation.

9) The only image color visible on the back side of the cloth is in the region of the hair. This probably can be expanded to include Giulio Fanti's measurements; however, he was not given definitive data by Turin.

10) No image formed under the blood stains.

11) The image-formation mechanism did not damage, denature, or char the blood. The blood can be removed with a proteolytic enzyme.

12) Image color can be chemically reduced with diimide, leaving colorless cellulose fibers. All image color resides on the outer surfaces of the fibers.

13) The medullas of colored image fibers are not colored: *The cellulose was not involved in color production.*

14) The color of image fibers was often stripped off of their surfaces, leaving molds of the fibers in the adhesive. Growth nodes can be seen in the molds. All of the color is on the surfaces of the fibers.

15) Chemical tests showed that there is no protein painting medium or protein-containing coating in image areas. It follows that microbiological activity did not produce the image.

16) Microchemical tests with iodine detected the presence of starch impurities on the surfaces of linen fibers from the Shroud.

17) There is no evidence for tissue breakdown (formation of liquid decomposition products of a body). Body fluids (other than blood) did not percolate into the cloth.

18) Any radiation that is energetic enough to cause the initial dehydration reactions of cellulose decomposition would penetrate into a fiber to a distance determined by its energy. The image fibers could not have been colored by energetic radiation.

19) Energetic radiation of all kinds causes defects in the cellulose crystals of the flax fibers. The defects are visible between crossed polarizers in a petrographic microscope. Shroud fibers show only normal aging.

20) Image fibers and non-image fibers show exactly the same kinds of defects and defect populations. The image was not caused by energetic radiation.

21) Rapid heating, as when linen is scorched with a torch, leaves characteristic, small balls of solidified melt at the ends of fibers. There are none on the Shroud.

22) The cloth does not show any phosphorescence.

23) The blood on the cloth is still largely red. Old blood is normally black.

Lateral neural inhibition:

Many observers look at the image for such a long time that they begin to see things that others do not. They attempt to use these observations to prove the resurrection of Jesus or some other belief.

The ability to see structure in amorphous bodies is responsible for our ability to see figures in clouds. Physiologically, the effect is explained in terms of "lateral neural inhibition": the human eye enhances edge contrasts. The mind plays games with what we think we see. Some devoted observers see images of flowers, teeth, bones, etc. on the Shroud. A statement like "I think I see" is totally unacceptable in a scientific discussion. These images are sometimes best seen after multiple contrast enhancements reduce the image to a pattern of dots.

EXAMPLES OF IMAGE-FORMATION HYPOTHESES THAT DEFY KNOWN FACTS:

The report that the Shroud dated to AD 1260-1390 has led to the formulation of many embarrassing "theories" (more correctly "postulates") simultaneously to explain both image formation and the unexpected age. Most of these have involved some form of "radiation," and most require a miracle to produce it. Nuclear radiation is required in these "theories" to produce excess radiocarbon and explain the date, while chemical reactions produced by the radiation are invoked to explain the image color. There are compelling scientific reasons why none of these "theories" can be accepted. All must be tested against the facts and observations.

Most of the hypotheses about the date and image have had a theological basis, and most claim that "this is the *only* way it could have happened." When the postulate fits the desired goal, little effort is made to test and reject it. The proponents then try to work from the postulated indispensable radiation to a proof of the Biblical resurrection. Many do not offer either a theoretical basis or experimental demonstration.

Goal-directed "theories" and pseudoscience have badly damaged the credibility of rigorous scientific studies on the Shroud of Turin.

Axions:
Some phenomena have been postulated without discussion. I have seen invocations of light, ultraviolet radiation, soft x rays, protons and other ionizing particles, neutrons, axions resonating in the cavity of the tomb and other perhaps not discovered forms of energy.

Scientists at the Livermore National Laboratory, MIT, the Lawrence Berkeley Laboratory, and Fermilab are currently working very hard just to detect axions, a very hard, expensive task. The *axion* is a hypothetical elementary particle proposed to explain the absence of an electrical dipole moment for the neutron. It has no electric charge, no spin, and would hardly interact with ordinary matter (electrons, photons, quarks, etc.) at all. It would be very unlikely to cause any chemical effects. Even though the axion -- if it

79

exists -- should have only a tiny mass, axions would theoretically have been produced abundantly in the Big Bang, and relic axions are a possible candidate for the dark matter in the universe. That is the reason they are being studied. They are not even a remote candidate for image production.

The dematerialized body:

Antonacci's "Historically Consistent Method" [found in *The Resurrection of the Shroud* by Mark Antonacci] is an outstanding example of goal-directed pseudoscience that involves radiation. He claims that it "was developed by combining research from scientists throughout the world on all aspects of the body images and blood marks on the Shroud...*This theory* (sic) *states that if a body instantaneously dematerialized or disappeared, particle radiation would be given off naturally and all the unique features found on the Shroud's body images and blood marks would occur*" (emphasis added). He claims that this is "the only way" the image could have formed. That is becoming the refrain of Shroud pseudoscience.

Figure X-1: One of Frei's fibers from an image area (200X), presented by Moran to prove that radiation-caused "pixels" of color along the fiber were "sharply terminated." It is misinterpreted; however, it clearly shows that the medullas are colorless.

The energy of nuclear weapons is based on the fact that E = mc^2; therefore, one bothersome problem with Antonacci's "theory" is that complete conversion of the mass of a normal human body into energy would have the effect of a huge H bomb, on the order of 200-300 **megatons** of TNT. However, particle radiation causes very distinct changes in flax fibers (see pp 89-93). Particle radiation absolutely was not involved in image formation. I believe he has done severe damage to the credibility of studies on the Shroud.

Optically Terminated Image Pixels:

Some observations can be misinterpreted and become quite mischievous. A good example appeared in unrefereed presentations by Kevin Moran as "Optically Terminated Image Pixels Observed on Frei 1978 Samples."

He says: "It is suggested that the image was formed when a high-energy particle struck the fiber and released radiation within the fiber at a speed greater than the local speed of light. Since the fiber acts as a light pipe, this energy moved out through the fiber until it encountered an optical discontinuity, then it slowed to the local speed of light and dispersed."

He bases his claim on a photomicrograph of an image fiber (figure X-1) from the collection that Max Frei took during the 1978 studies.

The light Moran postulates as responsible for color formation would have to be Cherenkov radiation. P. A. Cherenkov predicted the radiation in 1934, and it was explained theoretically by I. M. Frank and I. Y. Tamm. The three scientists shared the 1958 Nobel Prize for physics for the discovery. Cherenkov radiation is well understood.

In dielectric media the velocity of light is reduced to the value c/n (c = the velocity of light in a vacuum, and n = the index of refraction of the medium). Light goes slower in a dense medium than in a light one. If a charged particle hits a material at a velocity higher than the velocity of light in that material, Cherenkov radiation is emitted. Note that the particle can **not** go faster than the velocity of light in a vacuum. The Cherenkov radiation is simply light, and it travels through the medium at the velocity determined by the material's index of refraction, exactly the way all of the other light shining on the medium does. The radiation is emitted at an angle A to the direction of the particle's velocity, given by cos A = c/nv, where v is the particle's velocity. The wave front forms a conical surface carried along with the fast particle (similar to a shock wave in air), with the apex of the cone being the particle. The half-angle of the cone is π/2 - A. Moran did not address the amount of light that could be produced in a linen fiber, its direction of emission, or the wavelength of the light.

"I think I see" does not replace scientific study. The index of refraction of a transparent medium varies with the wavelength of light (dispersion). Moran seemed to think that the "radiation [light] within the fiber [was going] at a speed greater than the local speed of light." It could not.

Moran also assumes that the fiber can act as a "light pipe." It can not. Light pipes are made with a central fiber of a higher index of refraction than the outer part to keep the light inside. A linen fiber has a hollow core filled with air (low index of refraction) and nodes that look like microscopic bamboo joints.

Even the best transparent media absorb and scatter some light energy. Linen is not one of the best: It contains tiny rod-shaped cellulose crystals that are birefringent and scatter the light.

The intensity of light at any point in any medium can be predicted according to the Lambert/Bouguer law. The law says that the light absorbed divided by the incident light = -kd, where k is the absorption coefficient and d is the thickness. Each successive thickness behaves the same, with a reduced intensity of incident light.

Light-induced reactions fade as the light passes through an absorbing medium. The photomicrograph in figure X-1 shows a constant coloration along the fiber. The color was not a result of light-induced chemistry.

Moran postulates that the light will be dispersed at an "optical discontinuity"; however, the photomicrograph shows that color intensity remains constant past several "optical discontinuities," the nodes.

Moran's main mistake is a misinterpretation of Frei's photomicrograph. Linen fibers are *birefringent*; they have a different velocity of light along the fiber than across the fiber. When they are observed between crossed polarizers, they show light and different colors at different angles. Where there is no birefringence, the background is black. Figure X-2 shows the effect. The colors are totally a result of birefringence. The "sharply terminated pixel" in the middle of the field of view is the result of a change in the angle of the fiber *vis-à-vis* the angles of the polarized light beams used for observation. This is identical to the phenomenon shown with the Frei fiber in figure X-1.

Figure X-2: Modern, commercial flax fibers between crossed polarizers (400X) in 1.515-index immersion oil.

Figure X-2 shows light, birefringent nodes in the dark, vertical part of the fiber that has the abrupt change in color. That part of the fiber was turned to its angle of extinction, but the nodes remained birefringent. That proves that they are "optical discontinuities." The background of figure X-1 is lighter than that of figure X-2, because Frei's fiber was stuck to a piece of commercial transparent tape. Such tapes are birefringent. That is the reason we had special, amorphous tape made for taking the STURP samples.

When birefringent materials are placed one on top of the other, the birefringences add. The effect can be observed in the upper left quadrant of figure X-2 where two fibers cross at 90°. Colors shown in figure X-1 have been affected by the superposition of a fiber on birefringent tape. Figure X-1 was completely misinterpreted by Moran. His postulate had no scientific basis. A rigorous application of Scientific Method would have avoided the embarrassment to Shroud scientific studies.

The corona-discharge hypothesis for image formation:

Al Adler said that "The appearance of the image is consistent with a corona-discharge hypothesis." Although he did not explain how it is consistent, his statement spawned many claims and "theories"; however, in any scientific investigation, it is imperative to start with a clear, succinct statement of the hypothesis to be tested. I have not seen such a statement for this hypothesis, but I have culled the following from the literature and recent correspondence.

1) A voltage of sufficient magnitude to produce a corona discharge in air was produced in Jesus' body at some time while it was in the tomb and covered with the Shroud.
2) The corona discharge was not sufficiently intense to oxidize a significant amount of the cloth, and the potential difference was not great enough to produce sparks or arcs.
3) The corona "activated" the surface such that aging (and/or heating) produced a color.

Hypotheses that image color was a result of a corona discharge while the body and cloth were in contact must be tested against observations. Samples from the Shroud exist, and they can be observed by many methods. Corona discharges are a subset of plasmas, and discharges in the context of the Shroud would have to have occurred in air. Such discharges cause specific kinds of changes in flax fibers. The ionization potential of air is relatively low. It is quite easy for sparks and/or arcs to form. Sparks and arcs are very hot, but no spark or arc damage is seen on the Shroud.

The facts concerning corona discharges (plasma effects) can be found in physics texts. True corona discharges in air produce large amounts of oxygen atoms and lesser amounts of excited oxygen molecules. Both oxygen species oxidize flax efficiently, giving the same gaseous products as combustion in pure oxygen. Even small amounts of such a plasma will cause observable erosion and chemical changes on the surfaces of fibers. An oxygen-containing plasma *will* oxidize the material of the cloth. Such a process is used in commercial textile processing, making the textile easier to wet. Given enough time or intensity, a plasma in air will completely consume a linen sample. Even a short exposure will erode the surface of flax fibers. No such changes can be observed on Shroud fibers. Indeed, there is a very thin, discrete, colored coating of material on their surfaces.

A high voltage is required to ionize air and produce a corona discharge. A corona discharge is one type of plasma. One hypothesis is that the ionizing voltage appeared at the instant of resurrection, thus proving the event's miraculous nature. It has also been suggested that the voltage was produced by the piezoelectric effect during an earthquake, although electrical connections between the body and slip zone have never been elucidated. Whatever the voltage source, a corona discharge (or plasma) will produce specific effects on and in linen. If those effects are not observed, this image-formation hypothesis must be rejected.

In order for the body to charge to a high voltage, it must not be grounded. A body insulated from limestone only by a single thickness of moist linen should be at ground potential: no corona could form. In addition, body/cloth contact would bring the two surfaces to the same potential. Even small, light, non-conductive pith balls short out on contact when used to demonstrate electrostatics in elementary science classes. Because there is no such thing as a perfect insulator, materials in contact assume the same potential. Without a potential difference, ionization is impossible.

During a corona discharge, air is ionized in a high-voltage field to produce free electrons and positively-charged ions. The plasma is neutral; the number of electrons is the same as the number of charges on the positive ions. When electrons recombine with ions, light is emitted. Among other things, a corona discharge (plasma) produces energetic ultraviolet light. The predominant energy produced by a corona discharge that would be expected to produce high-free-energy zones in cellulose or impurities on its surface is its uv radiation. Colors form as a result of chemical reactions at high-free-energy zones.

The cellulose of image fibers is not colored; therefore, the ultraviolet light would have to produce high-free-energy defects in an impurity layer on the fibers' surfaces without affecting the cellulose. If no defects are in either an impurity layer or the components of the flax fibers, a coherent image can not appear on aging or heating. If radiation is intense, defects will appear in the cellulose. Defects in cellulose can be observed with a petrographic microscope.

Any corona in air will produce atomic oxygen and excited oxygen molecules: both

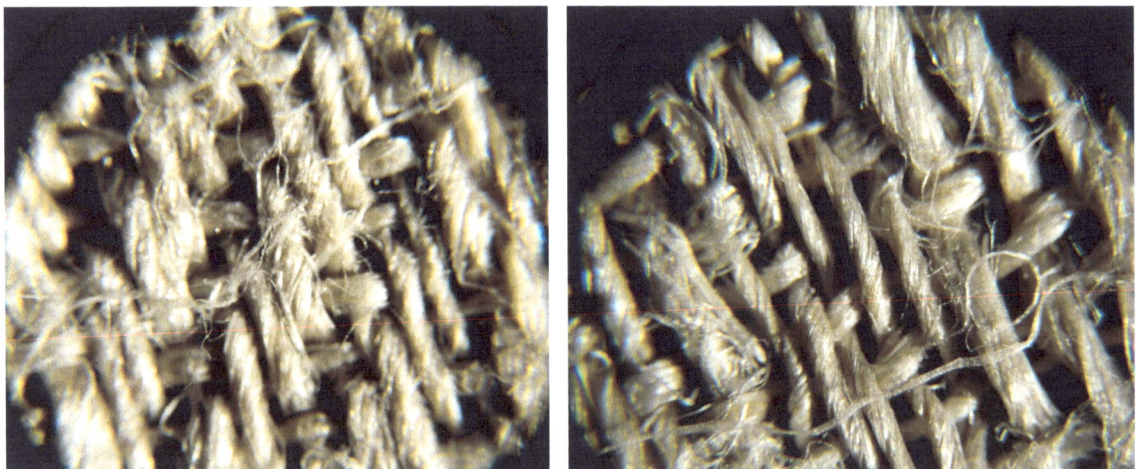

Figure X-3: (Left) Edgerton's linen before plasma treatment.

Figure X-4: (Right) Edgerton's linen after plasma treatment.

84

Figures X-3 to X-8 show plasma effects on linen surfaces. The experiments were run by APJet of Santa Fe, NM, a manufacturer of equipment for plasma treating textiles. The plasma was produced at 27 MHz in a 1% oxygen atmosphere. Conditions were modified from those in normal air to slow the rate of erosion of the linen; otherwise, the cloth would have been destroyed. The plasma nozzle was cooled, and the plasma temperature was only 85°C. Plasma energy was 220 W. The plasma jet was 6 mm in diameter, and the cloth was 4 mm from the tip of the nozzle. Exposure was 30 seconds, producing about 10^{19} free oxygen atoms. No sparks or arcs were formed. The sample was produced by Kate Edgerton (Norwalk, Connecticut, deceased), following the methods used in the Near East in Roman times.

It is easy to see that the nap of the cloth was removed in the area subjected to the plasma. The fibers were oxidized by the energetic oxygen atoms and molecules, even though the temperature was not high enough to dehydrate and color the cellulose.

Figures X-5 and X-6 show that the plasma completely penetrated the pores of the cloth. Because plasma is neutral, it does not charge the surface of an insulator, ultimately repelling itself, as does an electron beam. It penetrates the entire structure.

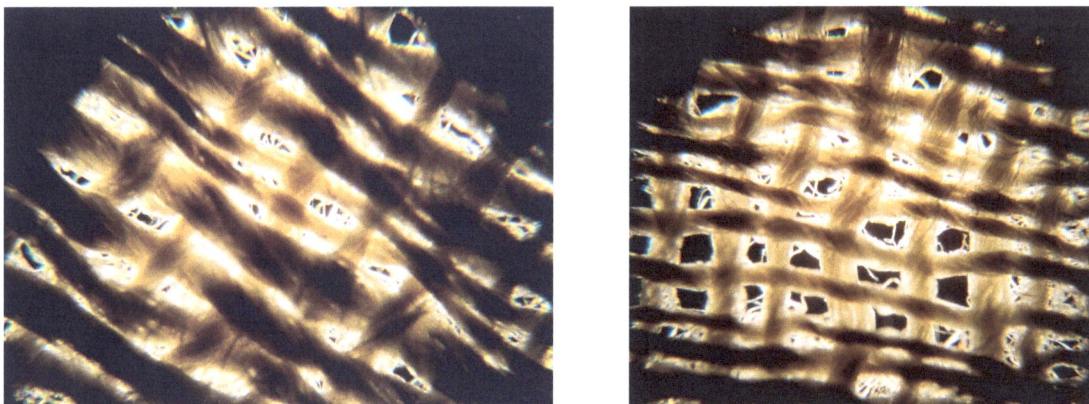

Figure X-5: (Left) The same sample as shown in figure 1. It is viewed in transmitted cross-polarized light. The pores are filled with fiber ends that have unwound from the yarn. The yarn has a Z twist.

Figure X-6: (Right) The same sample as figure 2, viewed in transmitted cross-polarized light. The fibers in the pores have mostly been oxidized. The plasma penetrated the pores.

Both the effects of surface erosion and the crystal defects caused by ultraviolet irradiation can be observed with a petrographic microscope (figures X-7 and X-8). It is clear that a corona discharge (plasma) in air will cause easily observable changes in a linen sample. No such effects can be observed in image fibers from the Shroud of Turin.

Any electromagnetic radiation more energetic than green light will produce chemical changes in flax fibers. These changes can be observed as diffuse birefringence throughout the cellulose crystals. Similar effects are caused by natural high-energy electromagnetic radiation; however, the Shroud provides its own internal standard for testing this effect. Image and non-image fibers show exactly the same amount of electromagnetic-radiation damage. The image did not receive excess radiation during its formation.

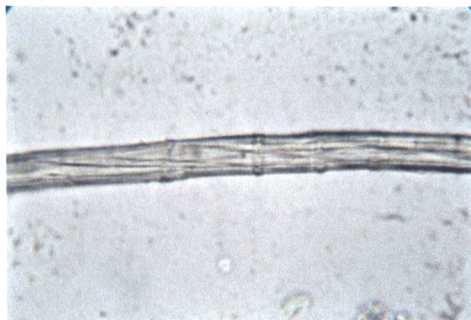

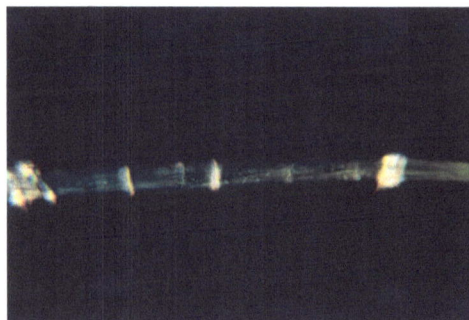

Figure X-7: (Left) A single fiber from the center of figure 2 in water. Hemicelluloses and pectins have been oxidized, leaving most of the more stable cellulose. The spiral structure of the cellulose ultimate cells is visible.

Figure X-8: (Right) The same fiber as figure 5 is shown between crossed polarizers. The bright birefringent spots are the growth nodes of the fiber. The white birefringent haze was caused by uv-induced defects in the cellulose crystals.

PROPERTIES OF ENERGETIC RADIATION:

Some general information is required to make valid tests of the many "radiation" hypotheses for image formation.

Nearly all instances of color formation are the results of some kind of chemical process. The only common exceptions are interference (e.g., a thin film of oil on water), diffraction (scattering from a system of parallel lines), and refraction (a prism or a rainbow) colors.

Chemical reactions can be initiated by different kinds of radiation. However, the different types of radiation are not well understood by the general public. There is widespread belief that radiation can cause some unexpected effects, but it is not magical. The properties of radiation have been exhaustively studied, especially since the advent of the "nuclear age." Radiation must not be invoked to explain the image without discussing its properties.

Jean-Baptiste Rinaudo claimed that, "Protons are indeed the only ionizing radiations to induce mainly acid oxidations." Protons have a positive electrostatic charge. Although the evidence indicates that the image color is mostly a result of complex conjugated carbon-carbon double bonds, a look at proton penetration data shows that his hypothesis does not and can not fit the observations (figure X-9). The penetration of forms of radiation into linen is critical in testing any hypothesis involving radiation. The National Institute of Standards and Technology (NIST) website, gives comprehensive information on penetration for different forms of radiation. An example for protons is

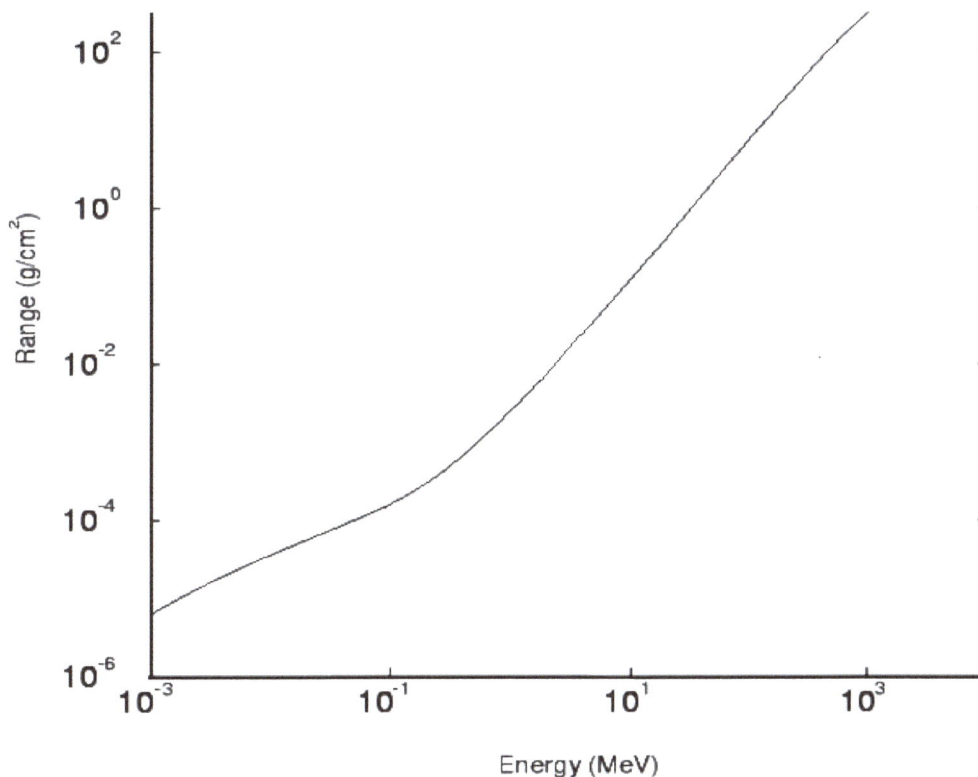

Figure X-9: NIST data for proton penetration as a function of energy.

shown in figure X-9. Some scientific background is required to interpret the data and graphs. It is easy to calculate the range of a proton in cm from the range in g/cm^2. For example, 1 cm^3 of water weighs 1 gram. A penetration of 1 cm would require roughly 16 MeV of energy (the energy scale is logarithmic). A proton with a few MeV energy would penetrate the entire diameter of a Shroud fiber, coloring the medulla on its way to the farther surface. Rinaudo used 1.4 MeV protons, which penetrated all of the way through the fibers. However, microscopy showed that the color he obtained was produced only in the layer of sizing on the surface of the new linen that was used for the experiments. Simple heating caused the color (figure X-10).

Figures X-11 and X-12 compare Rinaudo's proton-irradiated cloth with the image at the bridge of the nose on the Shroud. They have few if any similarities.

An average flax fiber is 10-20 μm in diameter, and some lower fibers are colored in image areas. Any radiation that colored the circumference of two, superimposed fibers would have to penetrate at least 20-40 μm of cellulose. Radiation that penetrated the entire 10-20-μm-diameter of a fiber would certainly affect the entire volume of the fiber, including the walls of the medulla (the cylindrical void in the center of the fiber). All image fibers show color on their outer surfaces, but the body of the fiber and the walls of the medulla are not colored.

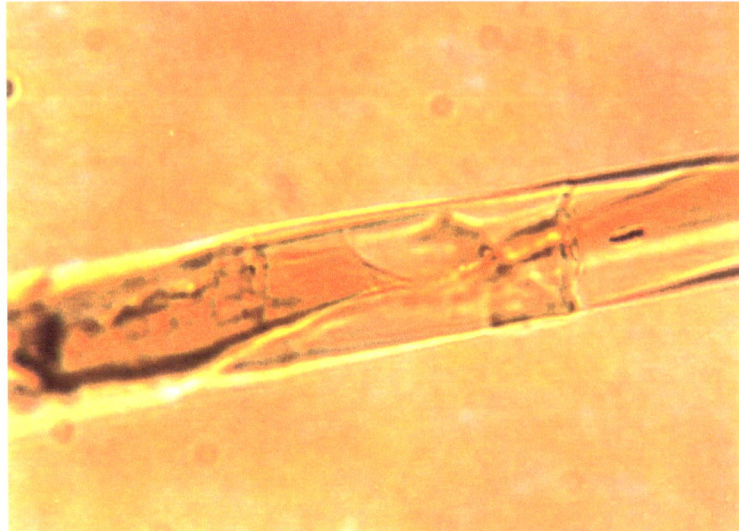

Figure X-10: One of Rinaudo's proton-irradiated fibers. The only color appears in the film of sizing material.

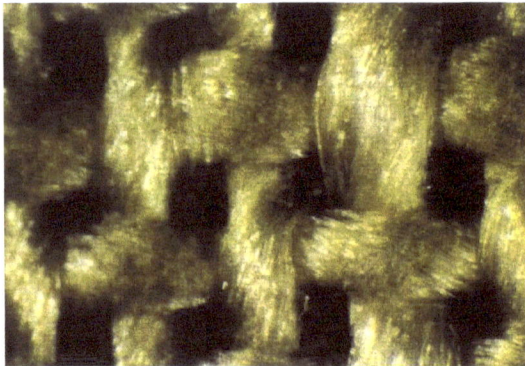

Figure X-11: (Left) Rinaudo's proton-irradiated linen.

Figure X-12: (Right) Dark image at bridge of nose on the Shroud. ©1978 Mark Evans

When energetic radiation of any kind penetrates an organic material like linen, it produces "defects." It breaks chemical bonds and moves things around. The crystals of the material are changed, and we can see those changes quite easily with a good petrographic microscope. Figure X-13 shows one of Rinaudo's proton-irradiated flax fibers between crossed polarizers in the microscope. This view can be compared with figure X-2, which had not been irradiated. The defects are all through Rinaudo's fiber, and there are many of them.

The protons ionized the cellulose as they passed through the fiber. This warped the crystals making the protons' paths birefringent. You can see where they went in the fiber by the straight lines of their paths. But not all of the effects of ionizing radiation appear instantly. Some of the chemical reactions at defects take time. We might not have been able to see all of the proton-produced defects immediately after irradiation. Defects also tend to coalesce into larger defects on standing. But sometimes it is nearly impossible to see the original defects, and they need to be "decorated" to become visible. They can be etched with chemicals, be dyed with some colored materials, or have their reactions accelerated by heating. However, the passage of ionizing radiation through linen fibers can always be detected. No such effects of ionizing radiation can be found in Shroud fibers.

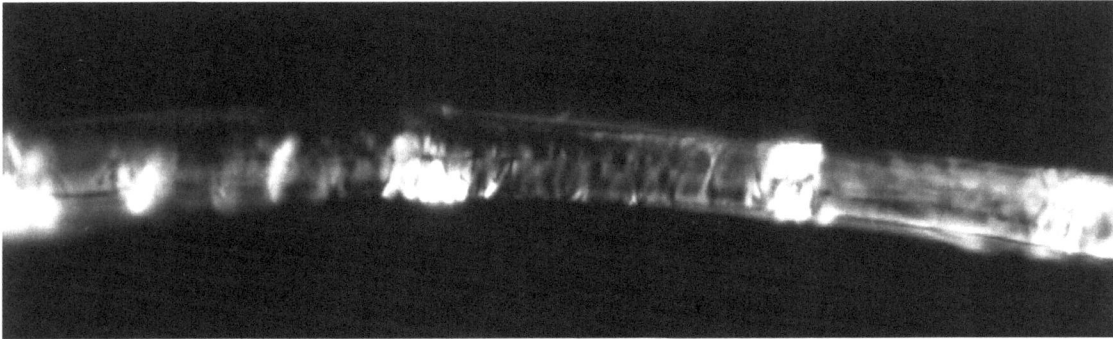

Figure X-13: One of Rinaudo's proton-irradiated fibers. Little, white, straight lines cutting across the fiber are the paths of the protons.

Not all kinds of radiation ionize the material they penetrate. Neutrons and neutrinos do not have any electrical charge. Neutrinos hardly interact with matter at all, the fact that made them so difficult to detect. They have practically no chance of being stopped as they shoot through the entire diameter of the earth. The effects of neutrons depend on their energy.

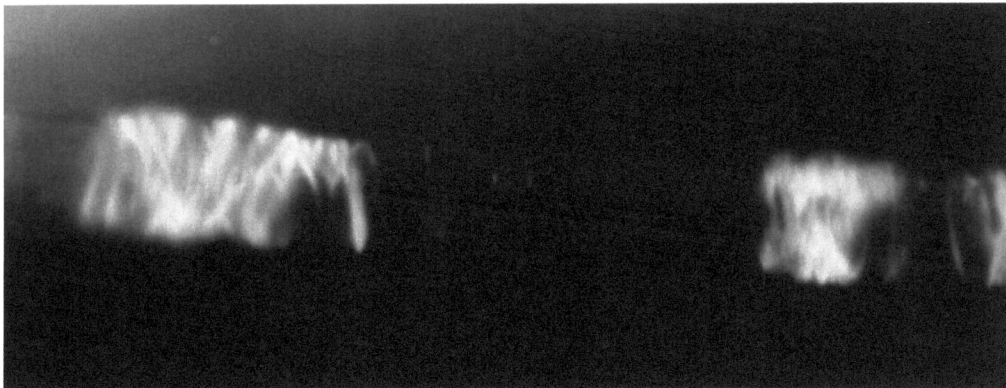

Figure X-14: One of Moroni's neutron-irradiated fibers from the Lyma mummy wrapping. Observe the small, white, vertical streaks between the bright growth nodes. There is also a faint haze in the background.

Mario Moroni did experiments with thermal neutrons from a reactor attempting to prove that the unexpected radiocarbon date was a result of that kind of radiation. Neutrons can convert nitrogen atoms into radiocarbon atoms. That is the way the ^{14}C is produced in the atmosphere. No other potential path to radiocarbon is anywhere nearly as probable.

One problem in making assumptions about linen is that it must contain a considerable amount of nitrogen. It usually does not.

Moroni got an apparent change in the age of an old mummy linen, but he did not report how much nitrogen was in his cloth samples. Corpses produce nitrogen compounds, but he did not work very hard to determine all of the details of his experiments. If you do not look for alternate reasons, you can claim a neutron-producing miracle.

A neutron weighs about the same as a proton. Protons are the cores of hydrogen atoms, and linen is loaded with hydrogen atoms. Slow, thermal neutrons tend to bounce around like tiny billiard balls in any systems that contain lots of hydrogen atoms. Occasionally, they hit head on with roughly the same effect as a direct hit between billiard balls: the one that was hit goes scooting off and the other sits in its place. When this happens between neutrons and hydrogen atoms, the result is a "recoil proton." That is a charged, ionizing particle. When you are working with neutrons, you can't see them in a cloud chamber, but you can see the tracks of the recoil protons. The same is true of linen. You don't see the track of a neutron, but you can see the tracks left by recoil protons. If you look very closely at figure X-14, you will see some little white streaks in the black. No type of radiation that could produce either color-producing chemical changes in the linen fibers or changes in the ^{14}C content could go unnoticed. All radiation has some kind of an effect on organic materials.

The mummy wrapping used by Moroni was quite old, but it did not show neutron effects (recoil protons) before it was irradiated (figure X-15).

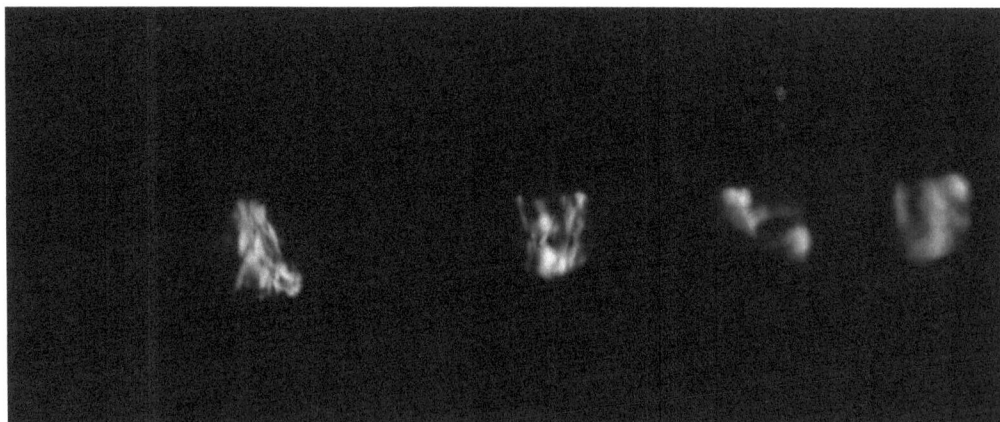

Figure X-15: A fiber from Moroni's Lyma-mummy wrapping before neutron irradiation. No recoil-proton tracks can be observed. There is practically no background haze.

CHEMICAL EFFECTS OF RADIATION:

When cellulose fibers are heated enough to color them, whether by conduction, convection, or radiation of any kind, the primary chemical reaction is the elimination of water from the structure (the cellulose is "dehydrated"). When water is eliminated, C-OH chemical bonds are broken. The C• free radicals formed are extremely reactive, and they will combine with any material in their vicinity. In cellulose, double bonds form between carbon atoms that have lost -H and -OH groups. As double bonds accumulate, color begins to appear. The reactive free radicals also recombine with other parts of the cellulose chains, and the chains *crosslink*.

A very large amount of information is available on radiation effects on materials. One important factor to know or measure is the "G-value" (the number of molecules destroyed per 100 eV of absorbed energy). Any claims made about radiation effects should be accompanied by enough measurements to support them. Even the crystalline, high-density, high explosives RDX and HMX have G values of only 2.9 and 1.4. A relatively large amount of energy is required to cause a significant effect in materials, and radiation produces physical effects as well as chemical effects.

All chemical reactions involve energy: Heat is either absorbed (endothermic reactions) or evolved (exothermic reactions). An explosion is an outstanding example of an exothermic reaction. ***One of the most important effects of radiation is to produce normal heating.*** Heat production can be calculated. Heat penetrates materials in known ways. Energetic radiation can not directly produce color without penetration, but sufficient radiation-induced heating can scorch linen and/or its impurities.

If radiation is not sufficiently energetic directly to cause chemical bonds to break, its only effect is to heat the material. Red and infrared light do not color linen unless they are so intense that they heat it to a temperature where the rate of dehydration is significant. The heat will penetrate the cloth.

The photographs of the back side of the cloth that were taken in June and July of 2002 show faint image color on the back of the cloth in the area of the hair. No body image is visible. What kind of radiation would penetrate the cloth and color it in the area of the hair and not penetrate the cloth anywhere else? Also, fibers taken from the face and hair images in 1978 are identical to all of the other image fibers: They are colored *only* on the surface. The cellulose was not colored. How can energetic radiation color just the outside of a very small fiber? Image fibers do ***not*** show the properties of scorched linen fibers.

The observations of colorless cores in image fibers, "ghosts" pulled from fibers by the adhesive, the reduction of the color with diimide, lack of fluorescence in an image area, and optical differences between image and scorch fibers eliminate any high-temperature heating event or energetic radiation in image formation. ***The cellulose of the colored image fibers has not changed as a result of image formation.***

Other than observing colored medullas, crystallinity and birefringence enable differentiating between scorched and image fibers. ***The evidence is strong that the image is not a result of dehydration of the cellulose fibers by any mechanism, thermal or radiation.***

Neutrons produce "recoil protons" when they hit a material that contains hydrogen. The loss of hydrogen also causes crosslinking and double-bond formation. Neutrons can not be invoked for image formation. Energetic neutrons would penetrate the entire diameter of the fibers.

The color of the image is indeed a result of a thin coating. "Thin" is the important word. Surface cracking ("corrosion" as Adler called it) of the color can be seen, and flakes can be seen in the "ghosts" on the sampling tapes (figure VII-2). It takes a thickness on the order of a wavelength of light to get an observable change in index of refraction, and observed indexes of an image fiber are identical to those of a fiber from the Holland cloth or modern linen. The image-color coating seems to be amorphous, but I have been unable to measure its index. I have been able to measure the index of the gum coating on the Raes sample. The thickness of the image color must be less than a sodium-D wavelength (589 nanometers).

Any photon (quantum of light) of a shorter wavelength than about 530 nm (more blue, more energetic) can cause the breakage of both C-OH and C-C bonds. The bond strengths of both C-OH bonds and a C-C bonds are about 90-91 kilocalories per mole (close to 3 ev). Less energetic light (more red) does not degrade cellulose. That is why clothing stores tend to use transparent orange curtains to protect the fabrics. Light that can color the entire circumference of a linen fiber will color the entire volume of the fiber.

John Jackson, one of the leaders of the STURP effort, has talked about "the physics of miracles." He has proposed the following theory (actually a hypothesis): 1) The body becomes "mechanically transparent [at the instant of resurrection], and the cloth falls into the body." 2) "Jesus' body became 'a body of light.' The light only penetrates air a millimeter or two ("if at all"); i.e., the air is opaque to the radiation." 3) "The cloth falling into the body is a transitional event, not instantaneous." 4) "Only the fibers on the cloth that were fully exposed in the energy field were imaged ... Deeper fibers were protected from the energy field by the fiber lying on top of them and therefore not imaged." 5) "When the cloth has fallen completely through the energy field, the fibers on the other side become exposed and are imaged by the energy field, except where they are protected or shaded by other fibers." 6) "The dorsal image is a contact image."

Robert L. Feller has written one of the best books on the factors that affect textiles [Robert L. Feller, Accelerated aging: photochemical and thermal aspects, The J. Paul Getty Trust, 1994, 292 pages]. He makes the following statement about the interactions between radiant energy (photons) and organic materials like linen: "...the primary step in photochemistry is the absorption of radiation followed by the dissipation of that energy through heat, emission of radiation (fluorescence or phosphorescence), transfer of the energy to another molecular entity, or the direct breaking of bonds." Light that does not penetrate air will be in the vacuum ultraviolet range, and that is what Jackson postulated. Such light will be very energetic. It will heat the cloth, and it will have a direct effect on the chemical

92

composition of the cloth. Any changes in chemical composition will change the birefringence of the flax fibers. No such effects are observed in the Shroud.

If the image were caused by any form of radiation, the structure of the flax fibers in the image areas must be significantly different from those in non-image areas. They are not.

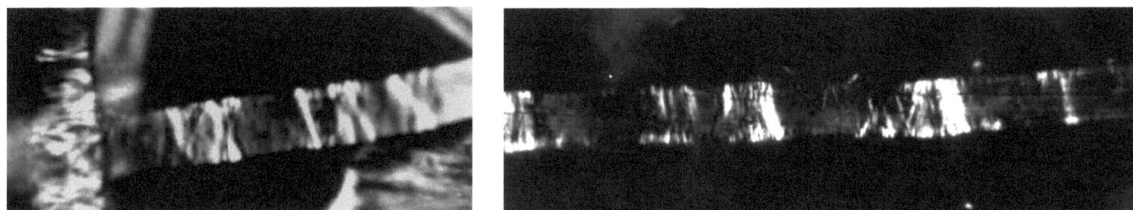

Figure X-16: (Left) A control sample from a non-image area. Its age is the same as any image area.

Figure X-17: (Right) A photomicrograph of an image fiber. The crystal damage is identical to that of the control, fig. X-16.

Figure X-16 shows STURP sample 3BF, a control sample that was taken in a non-image area to the left of the hand image. The photomicrograph was taken at 800X between crossed polarizers. There is a general light birefringence visible between growth nodes. This birefringence is a result of long-term, radiation-induced changes in the crystallinity of the cellulose. Because this is not an image area, the radiation had nothing to do with image formation.

Figure X-17 is a comparative view of an image fiber from STURP sample location 3EF. The same kind of light birefringence can be seen in it as is seen in the control sample. Many additional photomicrographs demonstrate this same fact. There is no significant difference in crystal perfection between image and non-image fibers. The image was not produced by radiation.

Conclusion on High-Energy Radiation:

High-energy radiation can not have been responsible for the color of the image. If it is sufficiently penetrating to color the entire surface of a fiber, it will color the entire volume of that fiber. That is not what is observed with image fibers. Energetic radiation of any kind will change the crystal structure of flax fibers. No such changes are observed in image fibers in the Shroud.

HYPOTHESES INVOLVING NATURAL PROCESSES:

Paul Vignon [in *The Shroud of Christ* (1902)] proposed what is called the "vaporographic theory" of image formation. He postulated that the body was covered with sweat at the time of death, and the sweat was rich in urea (N_2H_4CO). He also postulated that the cloth was saturated with aloetine (aloes extract in olive oil, an embalming liquid). The urea in the sweat would ferment to produce ammonia gas and carbon dioxide. The ammonia would diffuse into the cloth and react with the aloes to produce the image color. The hair image would be difficult to explain on this basis, but the distribution of image color and its lack of penetration into the cloth eliminated this hypothesis.

Sam Pellicori of STURP proposed a direct-contact hypothesis, and he did experiments to test it. He produced image-like colors, but color distribution was not correct. Sam coated his fingers with sweat, oil, and lemon juice and gently pressed them against cloth. Capillary flow could be observed. In order to get identical image colors over the entire cloth (including the hair), identical amounts of liquids would have to be distributed over the body. Assuming the weight of a body on the cloth, the identical properties of the front and back images seemed to eliminate direct contact.

STURP's Dee German modified the direct-contact hypothesis by assuming that the cloth sagged as a function of time and moisture, touching different areas at different times. This hypothesis could not explain the generally superficial nature of the image.

Eugenia Nitowski, an expert in Middle Eastern archaeology, pursued contact hypotheses farther than anyone else. She did experiments in ancient limestone tombs near Jerusalem. She experimented with a coated mannequin, and she tested slightly elevated temperatures (up to about 46°C). None of her results resembled the image.

Sebastiano Rodante experimented with a plaster head that had been soaked with bloody sweat. He soaked cloths in water and oily mixtures containing aloes and myrrh. He left the systems in contact for 36 hours, and he got images that were more realistic than Nitowski's. I have not seen photomicrographs of his images, and I would doubt that their structure was similar to the Shroud.

The extensive experiments seemed definitively to eliminate direct-contact hypotheses. None of the direct-contact systems could possibly have had the color penetration and distribution of the Shroud's image.

Many other image-formation hypotheses have been proposed, including Volkringer images (e.g., plants pressed in a book), singlet oxygen, and microbiological activity. When tested against the list of observations, none of the hypotheses could survive. I am not an expert on all of the different hypotheses. I encourage everyone to test their favorite hypotheses against the list of facts. I especially invite all readers to test my own hypotheses.

LOW-TEMPERATURE COLOR FORMATION:

Although high-temperatures and energetic radiation must be ruled out for image formation, lower-temperature processes are still possible. All that is required is that temperatures never reach the level where cellulose begins to dehydrate at a significant rate. Cellulose starts to dehydrate rapidly between 275 and 300°C. Many materials color rapidly at temperatures much lower than that.

94

Some claims have been made that thermal radiation (heat/infrared) could not play a part in image formation, because intensity of the radiation follows a $1/r^2$ law. That is not correct.

Materials radiate different wavelengths of electromagnetic energy at different temperatures. The wavelengths can be calculated from Planck's Radiation equation. Before a hot surface starts to glow, most of the radiation is produced in the infrared range. As shown in figure X-18, surfaces radiate different amounts of heat and light at different angles depending on the electronic structure of the material. Polished metals radiate and absorb very little thermal energy, and what they do radiate comes off of the surface at a very low angle. Nonmetals radiate much more thermal energy, and most of it comes out $90°$ to the surface. It is obvious that thermal radiation does not follow a $1/r^2$ law.

You can observe this effect by buffing the surface of a chrome-plated spatula and putting some non-metallic materials on it (e.g., a fingerprint, a spot of syrup, and/or a smear of clay). Heat the piece in the dark. As the temperature increases, the first places that you will see glow with radiation will be the print and other non-metallic spots, and they will glow long before the rest of the metal reddens with heat. The glow will be most visible from straight above. When the metal starts to glow, its light will be most visible at a low angle, looking along the surface.

The emissivity of a human body is like other non-metals or organic materials. Image formation that involves thermal radiation can not be ruled out; however, it can not explain all of the features of the Shroud's image.

The thermal conductivity of linen is low, about 2.1×10^{-4} cal cm^{-1} s^{-1} °C^{-1}; therefore, the temperature gradient extending outward from any heated area will be quite steep: It will be much hotter near contact points and cooler away from them. This is important in considering the chemical rates of processes that can form a color on a shroud that is in contact with a body.

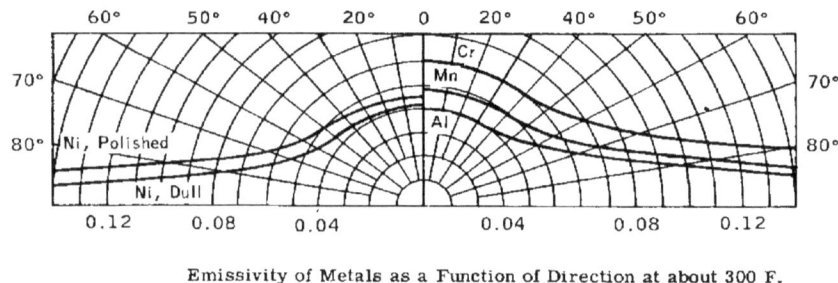

Emissivity of Metals as a Function of Direction at about 300 F.

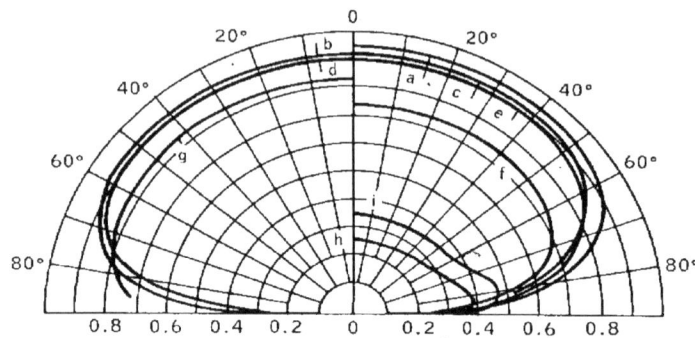

Emissivity of Non-Metals as a Function of Direction at 32-200 F. (a) Wet Ice. (b) Wood.
(c) Glass. (d) Paper. (e) Clay. (f) Copper Oxide. (g) Aluminum Oxide. (h) Aluminum Paint. (i) Bismuth

Figure X-18: Angular dependence of energy emission from different types of surfaces.

Chemistry and rate processes:

All chemical processes occur at some rate at every temperature above absolute zero, -273°C (-459°F), but the rates follow an exponential function as temperatures change. Some rates are very slow at normal temperatures. Cellulose does not scorch at a visible rate at room temperature, but it will char quickly above about 300°C. Other carbohydrates (e.g., starches and sugars) can color at much lower temperatures. A piece of cloth heated to several hundred degrees at a point will start to color, but the color quickly becomes less intense away from the heated point as a result of the low thermal conductivity of cellulose. Think of the scorch pattern a hot iron makes on a white shirt.

Several medical investigators have told me that the postmortem body temperature of a person who has died of hyperthermia and/or dehydration often reaches 41°C (106°F), and body temperatures can actually increase slightly after death. Some bodies have shown temperatures as high as 43°C (110°F). The bodies are quite dry as a result of the hyperthermia. Many bodies are found at high temperature in closed automobiles or lying on the desert in the sun. Materials that are stable at normal room temperatures (about 22°C) can react rapidly at 41-43°C. It should be noted, however, that bodies cool at a significant rate when they are in a cool environment.

96

Many simple chemical processes double (or even triple) their rates for each 10°C (18°F) increase in temperature. A dehydration-type of reaction could be expected to be about three times faster at normal body temperature (about 37°C) than at room temperature and *four to nine* times faster at about 41°C.

Chemical rates are modeled with an exponential equation called the Arrhenius expression, $k = Ze^{-E/RT}$, and rates can be predicted from known, measured chemical kinetics constants (k, the rate constant; Z, the frequency factor; E, the activation energy; R, the gas constant; and T, the absolute temperature). *Any chemical process involved in image formation will have properties in accordance with this equation.* All color formation processes involve chemical reactions.

Heat is also transferred by convection. The circulation of air between a hot surface and a cooler one is driven by the differences in density between a hot gas and a cooler gas. Convection cells are small where clearances are small and larger where clearances are larger. Convection also transfers vapors, which can include reactive gases, from one surface to another.

Fairly thin stagnant zones of gas form near fixed surfaces. Other gases that approach such zones must *diffuse* through the stagnant gas to reach the surface. Diffusion of gases through other gases is modeled with Graham's Law of Diffusion,

$$\frac{v_1}{v_2} = \frac{\sqrt{d_2}}{\sqrt{d_1}}$$

where v = velocity of diffusion and d = the density (or molecular weight) of the gas. The rates of diffusion are inversely proportional to the square roots of the densities of the gases. Diffusion parallel to the surface of a cloth that covers a body can not be instantaneous, and it will be slower for heavier molecules.

In the context of image-formation hypotheses that involve reactive gases, remember that cloth is porous. Gases diffusing to the cloth surface can pass through the pores and be lost. This phenomenon will cause a concentration gradient in the reactive gases. Color densities will be higher close to the body. Image-color density will be a function of body-cloth distance.

Cloth surfaces are active and adsorb gases rapidly, a fact that further limits concentrations as a function of distance and increases color density close to a body.

John Jackson's mathematical analysis of image resolution in the *1977 United States Conference of Research on The Shroud of Turin* suggested that no single, simple molecular-diffusion or radiation mechanism could produce the image observed. However, a combination of systems could offer an explanation, e.g., limited gaseous diffusion rates, exponential temperature-dependent chemical rate processes, anisotropic heat flow by radiation from the body to the cloth, attenuated heat-flow in the cloth, convection, surface properties of cloth, and diffusion through cloth pores.

Summary:

Image spectra suggest that some type of carbohydrate dehydration reaction would be most probable as an explanation for the image color; however, *the color appears only on the surface of individual fibers.* High temperatures and energetic radiation absolutely can **not** explain the properties of the image. *That statement does not suggest a miracle.*

CHAPTER XI: A COMPLEX, NATURAL IMAGE-FORMATION HYPOTHESIS

The fact that image color resides only on the fiber surfaces leads to the hypothesis that the color formed as a result of chemical reactions involving impurities on the surface. The spectra strongly suggest that the impurities were carbohydrates that dehydrated as a result of the image-formation process. The hypothesis on carbohydrate impurities is supported by observations of traces of some starch fractions on image fibers. No protein impurities were found in image areas.

Although scorched fibers show darkened medullas, the cellulose of image fibers is not colored. This proves that any impurities that produced the color had to have undergone low-temperature chemical reactions.

Cellulose dehydration follows known chemical rate laws. Cellulose is a very large polymeric chain made from glucose (sugar) units that are linked glycosidically. The glucose units exist in the "pyranose" form, rings containing five carbon atoms and one oxygen atom. Pyranose structures appear to be much more stable than are the other common sugar structure, furanose rings. Pyranose systems seem to have an activation energy for dehydration of about 27 kcal/mole, while furanose systems seem to be much less stable with an activation energy of about 19 kcal/mol. Because chemical rates are exponential with temperature, cellulose would react much more slowly than other carbohydrates at the same temperature.

In developing an hypothesis, I had to search for low-temperature chemical processes that produced the observed type of color. The processes had to involve only impurities. I had to eliminate direct "caramelization" of carbohydrates, because required temperatures were too high. Low-temperature processes must involve reactants other than those found or postulated on the cloth. Some reactive foreign material had to have come into contact with the impurities on the Shroud.

Some clues to possible low-temperature reactions might be obtained from the history of linen technology before about the 16th Century.

The bleaching methods used after about AD 1200 did not leave significant amounts of carbohydrate impurities. Pliny the Elder's history from about AD 77 suggested that we could expect traces of all of the carbohydrates that are found in crude starch on a piece of ancient linen. The warp yarn was protected with starch during the weaving process, making the cloth stiff. The final cloth was washed with "struthium," *Saponaria officinalis*, to make it more supple. In addition to starch fractions, we might expect traces of the glycoside sugars from *Saponaria officinalis* (e.g., galactose, glucose, arabinose, xylose, fucose, rhamnose, and glucuronic acid).

Dehydration-rate calculations made direct dehydration of any carbohydrate impurities, even pentose sugars, seem very unlikely. Sustained temperatures on the order of 100°C would be required to give a color in any reasonable amount of time (less than months).

A person hoaxing the Shroud image could not have lived long enough to produce a room-temperature image. Using a cloth washed in a low-surface-tension solution containing pentose sugars (e.g., a *Saponaria* solution) and dried in the sun, a non-metallic statue at a temperature above 100°C might have worked. Such an hypothesis could be tested; however, using Occam's Razor, another hypothesis seemed more probable.

A strong candidate for a low-temperature system that could produce the image color is what is called Maillard reactions. Maillard reactions are familiar to everyone who has seen bread bake, enjoyed dark beer, or had roasted coffee. The reactions are extremely complex, fast, color-producing reactions that involve several reaction paths and products. Starting with a complex mixture of carbohydrates, there could be thousands of products. Just one example of a product, the most intensely colored from reactions between xylose and L-alanine, is shown in figure XI-1.

A fascinating requirement for a hypothesis involving Maillard reactions is that amino groups ($-NH_2$) are needed to react with the carbohydrates. The potential source of amines, a decomposing body, involves support for the hypothesis that the Shroud is a real shroud.

The Maillard reactions are some of the most studied reactions in food chemistry. The reactions are commercially important, and they have been studied for many years. They involve the condensation of amino groups (the $-NH_2$ groups in amines, proteins, peptides, etc.) with reducing sugars and reducing polysaccharides such as sugars and some starch fractions. All reducing sugars and polysaccharides take part in the reactions. When amines and reducing sugars come together, *they will react. They will produce a color*. This is not a hypothesis: this is a fact. A cloth with crude starch on it *will* ultimately produce a color, if it is left in close proximity to a decomposing body.

The reactions occur at significant rates at much lower temperatures than the caramelization (thermal dehydration) of any of the sugars. A good example of Maillard reactions is the production of dark beers at low temperatures by reactions between maltose and any reducing starch components and the proteins or amino acids in the wort. Cheaper beers may be colored more rapidly by heating the wort with ammonia (the simplest amine).

The first steps of the Maillard reactions are rather fast at normal temperatures, and they produce colorless compounds (for example, glycosylated-proteins). The rates are even higher at body temperatures; however, they increase by factors between two and three for each 10°C (18°F) increase in temperature.

The colorless compounds are unstable, and they rearrange to give brown polymeric materials, melanoidins (figure XI-1), most of whose structures are still unknown. It takes

100

some time at lower temperatures for the color to appear. The color is not a result of oxidation.

Browning rates depend on pH (the acidity of the system). Rates peak in the neutral to basic pH range, but rates are still rapid in the 3-7 pH range. Human sebaceous secretions from the skin are about 28% free fatty acids; therefore, human skin is normally slightly acid, but Maillard reactions are still fast.

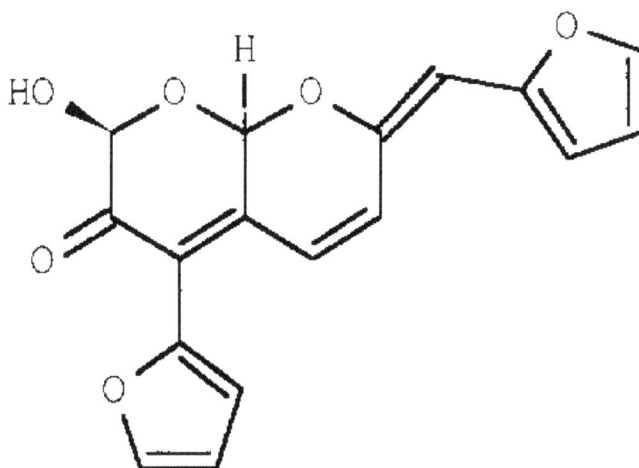

Figure XI-1: Example of an intensely colored Maillard reaction product. The color is a result of conjugated double bonds. The product does not contain any nitrogen.

Many of the final products of Maillard reactions are identical to those produced by caramelization of sugars. The structures that produce the color are conjugated double bonds, just as hypothesized from the spectra taken by STURP. Some of the most important products in color formation do not contain any nitrogen. This fact could help explain why we did not observe any nitrogen compounds in image areas.

With regard to the requirement for amine reactants in Maillard processes, decomposing bodies start producing ammonia and amines, e.g., cadaverine (1,5-diaminopentane) and putrescine (1,4-diaminobutane), fairly quickly, depending on the temperature and humidity. These compounds are responsible for the odor that develops around a body some time after death.

Most of the very volatile ammonia diffuses out through the nose and mouth soon after death. This fact may explain the darker image color between the nose and mouth and the penetration of image color in the vicinity of the hair.

The ammonia and many of the decomposition amines are volatile and basic (they increase the pH into a more favorable range for Maillard reactions), and they rapidly undergo Maillard reactions with any reducing saccharides they contact. The reactions are rapid at room temperature, or even lower. Such sugar-amine reactions offer a natural explanation for the color on the Shroud, *and they suggest that the Shroud of Turin was a real shroud.*

However, identification of a probable chemical process does not explain one of the most perplexing observations on the Shroud, the discontinuous distribution of the color on the topmost parts of the weave.

The fact that color does not penetrate the cloth to any significant depth suggests that any hypothesized impurity had to be concentrated at the top of the cloth with little throughout the rest of the thickness. Assumed pressure of a body had no effect on the color density or penetration of image color into the cloth in the area of the back image. This supports the hypothesis that a limited amount of superficial impurities was involved in image formation. There is a simple explanation.

Evaporation concentration was demonstrated in figure VI-3. It can explain how most of the color-producing impurities were concentrated on the upper surface of the cloth. The faint image of the hair on the back of the Shroud indicates that some impurities appeared on the back, as they do in most of my experiments.

Saponaria officinalis, also called "soapweed," reduces the surface tension of water making it a good wetting agent, a "detergent." Both hydrophobic and hydrophilic impurities on cloth are put into solution or suspension by *Saponaria.* The process is identical to that involved on every laundry day with every kind of soap or detergent.

We have now discussed critical observations on the Shroud, and we can clearly state an alternate hypothesis in accordance with Scientific Method.

Formal statement of an impurity hypothesis for image formation to be tested:
The cloth was produced by technology in use before the advent of large-scale bleaching. Each hank of yarn used in weaving was bleached individually. The warp yarns were protected and lubricated during weaving with an unpurified starch paste. The finished cloth was washed in *Saponaria officinalis* and laid out to dry. Starch fractions, linen impurities, and *Saponaria* residues concentrated at the evaporating surface. The cloth was used to wrap a dead body. Ammonia and other volatile amine decomposition products reacted rapidly with reducing saccharides on the cloth in Maillard reactions. The cloth was removed from the body before liquid decomposition products appeared. The color developed slowly as Maillard compounds decomposed into final colored compounds.

This hypothesis must be tested against all natural laws and observations.

Test and confirm:

It is not practical for an individual to attempt a complete reproduction of the Shroud; however, several components of the hypothesis can be tested independently. Many complex scientific hypotheses (e.g., Einstein's Theory of General Relativity) must be tested this way, one component at a time. Any person interested is free to test any component of this hypothesis, and I would applaud a complete test. I will not volunteer my body for the test. I have already used my blood to test effects of the fire.

Both starch and cellulose are composed of glucose units; however, cellulose is not a reducing polysaccharide. Crude starch contains a complex mixture of molecular weights as well as some very soluble free sugars: Its reducing carbohydrates rapidly undergo Maillard reactions.

Age and/or pyrolysis products and heating during the fire of AD 1532 have changed the lignin in the cloth, and it would certainly change other impurities that are less stable than either lignin or cellulose. Therefore, tests of image-formation hypotheses need fresh material. However, modern linen has been vigorously bleached, and it has been coated with sizing compounds and fluorescent fabric brighteners. It is useless for experimental image-production purposes.

Kate Edgerton (deceased, Norwich, CT) grew flax and made some linen, using both starch and *Saponaria*. Unfortunately, she ironed the samples with a "warm iron" which colored the cloth and changed the *Saponaria* glycosides. Edgerton's samples could be bleached to remove most of the color by soaking in 3.5% hydrogen peroxide. This treatment did not remove the natural waxes.

Because Edgerton's linen was scarce, many preliminary tests were performed on pure-cellulose, "ashless" filter paper. These experiments were done by placing drops of test solutions on a dry plastic plate and laying a piece of filter paper over them. The liquid migrated through the paper and evaporated at the surface. No color could be observed in either sunlight or ultraviolet illumination with *Saponaria* alone or with any model saccharides (e.g., levulose, dextrose, maltose, lactose, xylose, dextrin, and soluble starch).

The different samples were treated with ammonia vapor for different times. Light colors developed slowly on the tops of samples that had contained reducing saccharides.

A technical grade of dextrin was used to model crude starch. It acts like crude starch without the free sugars and highest molecular-weight fractions, and it reduces Fehling's solution (it is a "reducing" polysaccharide).

The Fehling's test uses a fresh mixture of copper sulfate and alkaline tartrate solutions. Add the sample, and heat gently. A reddish precipitate of cuprous oxide proves a reducing saccharide.

Paper was laid over different numbers of drops of dextrin, and the liquid was allowed to migrate into circular spots and evaporate. Samples were then laid over drops of *Saponaria* solution. The *Saponaria* caused the polysaccharides to move radially, as in a circular paper chromatogram. Treatment with ammonia showed that the Maillard colors were most intense where the dextrin had been concentrated by migration at the *Saponaria*-solution front and on the paper's surface.

Very little color was obtained when the same experiments were repeated with a purified, "soluble" starch or plant gum. The starch gave a bright blue color with iodine and it showed only the slightest reaction with Fehling's solution. The plant gums did not show reducing properties. The effects would have been different after hydrolysis of the materials.

It became evident that image-like colors required both saccharides and amines. The demonstration shown in figures XI-4, 5, and 6 were made with Edgerton's linen to reproduce effects with a primitive woven fabric.

Figure XI-4: Evaporation surface of "primitive-type" linen treated first with dextrin then *S. officinalis* solution. The Maillard color was developed with ammonia vapor and heating.

A sample of Edgerton's bleached linen was placed on four drops of dextrin solution on a plastic plate. A round spot was obtained. The water was allowed to evaporate from the cloth. No color could be seen on either surface. The middle of the same sample was placed on four drops of *Saponaria* solution. The wet spot expanded radially through the cloth. The water was allowed to evaporate, and no color could be observed. The sample was then treated at room temperature for 10 minutes with ammonia vapor. A very light color could be observed on the top surface after standing 24 hours at room temperature.

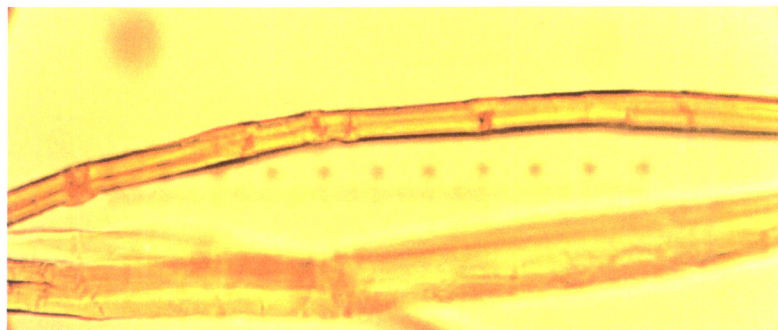

Figure XI-5: Maillard-colored fibers (400X) from the top of the sample in figure XI-4. Medullas are clear.

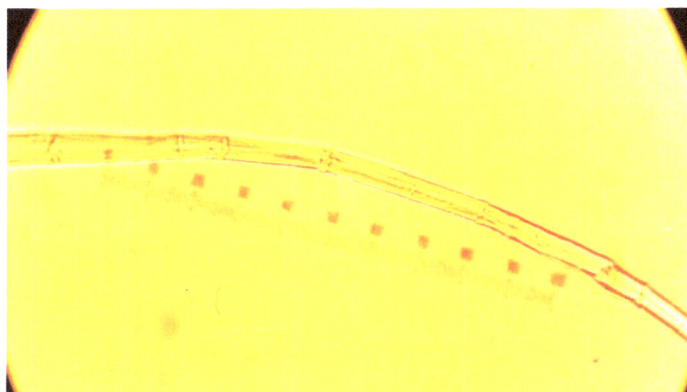

Figure XI-6: A nearly-colorless fiber from the middle of the back surface of figure XI-4. Reactants were concentrated at the evaporating surface.

The same effects as aging can be obtained by heating Maillard reactions. The reaction rates are determined by the Arrhenius activation energies and pre-exponentials of the processes, and the rate is an exponential function of the temperature. Reactions that require years at normal temperatures can be obtained in minutes at elevated temperatures. The rates of different Maillard reactions increase by factors between two and three for each 10°C (18°F) increase in temperature. This indicates that they have relatively low activation energies; rates will be significant at lower temperatures. The color shown in figure XI-4 developed after a few minutes of heating at 66°C (150°F). This temperature is far too low to scorch cellulose.

Figure XI-4 shows that the most intense color appears in the ring on top. Some color appears around the ring on the back surface; however, the center of the back is nearly colorless.

Experimental manipulations of concentrations and one-dimensional migration of solutions, as in a large cloth, could produce the same front-to-back color separation and color density as observed on the Shroud.

The fibers on the top-most surface are the most colored when observed under a microscope, and the color is a golden yellow similar to that on the Shroud (figure XI-5). The coating of Maillard products is too thin to be resolved with a light microscope, and it is all on the outside of the fibers. There is no coloration in the medullas: The color formed without scorching the cellulose.

There is very little color on fibers from the middle of the back surface (figure XI-6). The color-producing saccharides had concentrated on the evaporating surface.

Water-stained image areas on the Shroud showed that image color does not dissolve or migrate in water. Maillard products are not water soluble, and they do not move when wetted.

As a peripheral, non-scientific comment, several Shroud researchers have wondered why there is no mention of an image on the "cloths" reportedly found in Jesus' tomb. Assuming historical validity in the accounts, such a situation could be explained by the delay in the development of the Maillard reactions' colors at moderate temperatures. No miracle would be required.

Other considerations from physical chemistry:
The chemistry of the color does not answer all questions about how the "photographic" image formed. The image seems to show the body of a man, and it is darkest in areas that should have been closest to the body's surface; however, the "resolution" of the image has been puzzling. I believe that its resolution is a natural consequence of the image-formation process.

Vapor diffusion parallel to the cloth's inner surface would follow Graham's Law of diffusion, and it would be slow for high-molecular-weight decomposition amines. High Maillard reaction rates would further limit the spread of reactive amine vapors. Image densities would fall off rapidly away from the body, increasing resolution. Gaseous reactive amines can be lost by diffusion through the porous cloth, reducing concentrations and reaction rates inside the cloth. However, it has long been recognized that the images of the hair, moustache, and beard are anomalous. Figure XI-7 shows a slightly contrast-enhanced view of the area of the face and hair. The density of the image is greatest in those areas. That can easily be explained by the inhibition of vapor diffusion through a porous mat of hair. Ammonia is first evolved from the lungs; therefore, its concentrations would have been highest in the vicinity of the nose and mouth, and its diffusion would have been retarded by the moustache and beard. By the time heavy decomposition amines appear, the body will have cooled. The surface area of cloth is large and higher-molecular-weight decomposition amines adsorb strongly. All of these phenomena would cause a rapid reduction in amine concentrations away from contact points and the nose-mouth area.

Postmortem body temperatures can reach 43°C (110°F), and steep temperature gradients would exist across the cloth as a result of the low thermal diffusivity of linen and the angular dependence of radiant heat flow from a nonmetallic surface. The temperature gradients will have a large effect on Maillard reaction rates and image resolution before the body cools, i.e., while ammonia is the predominant amine.

Figure XI-7: Slightly enhanced view of image face. Note increased density in the nose-mouth and hair areas. The back side of the cloth shows the hair, it may show part of the face, but it does not show any body image. © 1978 R. N. Rogers

I believe that a complex combination of chemical and physical factors could produce a distribution of reaction products with the resolution of the image; however, the cloth would have to be removed from the body before liquid decay products appeared.

This complex system is a testable hypothesis; however, the many variables would require extensive individual experimentation.

Figure XI-8 shows one experiment designed to test the factors affecting resolution. A papier maché hand was filled with rice to increase its heat capacity, and it was heated to about 43°C. A piece of repeatedly-washed muslin was saturated with a dextrin solution, and it was washed in *Saponaria officinalis* suds. The heated hand was quickly painted with a 1:1 solution of diethylenetriamine (DETA) in acetone, and the acetone was allowed to evaporate. The cloth was laid over the hand, and the assembly was left in still room air (22-

23°C) for 12 hours. Practically no color change could be observed on the cloth when it was removed, but the color developed quickly on heating to 65°C. A more-quantitative test of resolution can be seen in chapter XII.

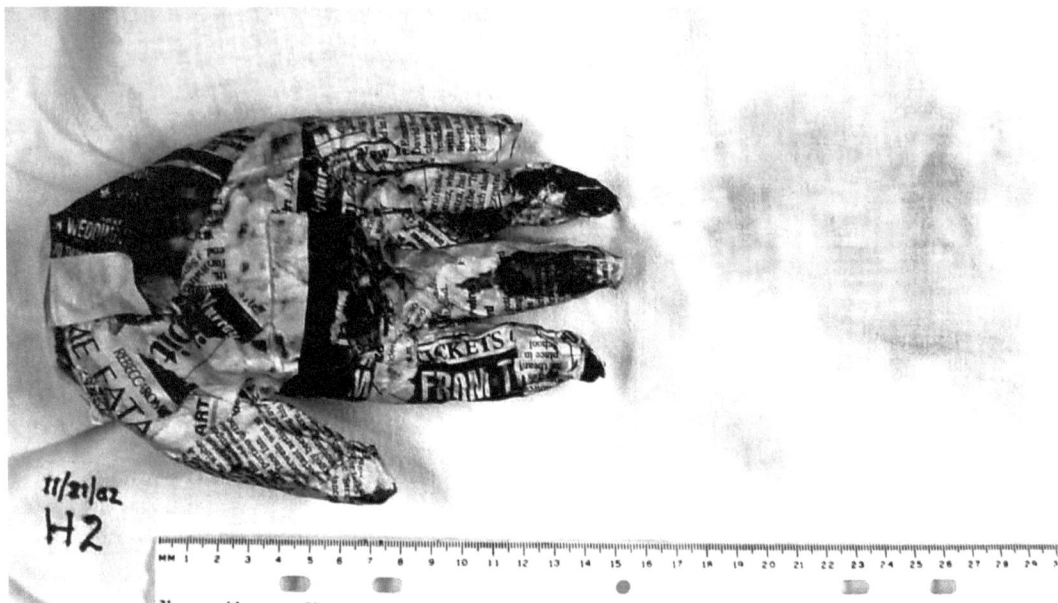

Figure IX-8: Image-formation experiment with a heated papier maché hand. Convection decreased resolution, and the amine vaporized too quickly; however, the thumb and one finger are clearly resolved.

The amine/saccharide experiments showed that the following variables are important: 1) When the "body" temperature is too high, convection cells[1] are too active, diffusing amines too widely for good resolution. Resolution improves at lower temperatures. A body that had cooled for several hours but has not yet produced high concentrations of amines would give better resolution than a hot body. 2) The amines must be released *slowly*. Too much amine badly reduced resolution. A decaying body would give much better resolution than any object that had been painted with pure amines. Too much amine would color the entire cloth, obliterating the image. A successful image that involved a real body would require removal of the cloth before extensive decomposition. 3) The experimental assembly must be kept in a space that is cool and still. 4) An increase in the concentration of reducing saccharides (impurities) on the cloth improves resolution. 5) Modern linen that does not contain suitable impurities will not produce an image.

Assuming the amine-impurity image-formation mechanism, the image on the Shroud required just exactly the correct conditions, or it would not have been produced with the resolution and color density observed.

108

SUMMARY:

A summary of the observations that I believe can now be confirmed and that must be used in hypothesis testing follows.

1) **No painting pigments or media scorched in image areas or was rendered water soluble at the time of the AD 1532 fire.** Different areas of the Shroud show intersections of scorches with image and blood. Some scorched areas had been wetted, and water percolated through the cloth by capillary flow. Nothing moved from an image area when water flowed through it.

2) **Direct microscopy showed that the image resided *only* on the topmost fibers at the highest parts of the weave.**

3) **The color density of any specific image area depends on the batch of yarn that was used in its weave.** Bands of different color can be observed, and they correspond to the batches of yarn that were used to weave the cloth. Image density is greater where darker bands cross, and it is lighter where lighter bands cross. This suggests that significant variations in impurity concentrations existed among yarn batches.

4) **Adhesive-tape samples show that the image is a result of concentrations of yellow fibers.**

5) **The color of image fibers was often stripped off of their surfaces, leaving molds of the fibers in the adhesive.** The molds show both growth nodes and image color. The color in the molds has the same chemical characteristics as the image color. The color resides only on the surface of the fibers. The layers of color are extremely thin.

6) **Reflectance spectra, chemical tests, laser-microprobe Raman spectra, pyrolysis mass spectrometry, and x-ray fluorescence all show that the image is not painted with any of the expected, historically-documented pigments,** including iron oxides. The image spectra were essentially identical to those from aged linen and light scorches. The structures of all forms of dehydrated carbohydrates would be very similar, containing complex systems of conjugated double carbon bonds. Cellulose is not unique. Sugars and starches give the same types of dehydration/conjugation chemical structures. Identical colored structures are produced by low-temperature reactions between reducing carbohydrates and amines, i.e., Maillard reactions.

7) **Chemical tests showed that there is no protein painting medium or protein-containing coating in image areas.** The image was not painted with glair (egg white), and there is no significant amount of microbiological coating on the cloth. Both McCrone's hypothesis that the image was painted with glair and hematite and Garza-Valdes' hypothesis that it was a result of microbiological activity can be rejected.

8) **Image color can be chemically reduced with diimide, leaving colorless cellulose fibers.** The color resides only on the surface of the fibers, and it is the result of conjugated double bonds. The underlying cellulose (linen) fibers are not colored.

9) **The image of the dorsal side of the body shows the same color density and distribution as the ventral, and it does not penetrate the cloth any more deeply than the image of the ventral side of the body.**

10) **The image does not fluoresce under ultraviolet illumination.** Scorch margins from the fire of AD 1532 fluoresce. The image was not caused by scorching, intense heating, flash heating, flash photolysis, ionizing radiation, or any other process that would

produce second-generation, fluorescent, chemical-decomposition products. Image color formed under mild conditions.

11) **Microchemical tests with iodine detected the presence of starch impurities on the surfaces of linen fibers from the Shroud.** Impurities were detected that could take part in color-producing Maillard reactions.

12) **The medullas of colored image fibers are not colored:** *The cellulose was not involved in the color-producing chemistry of the image.* Fibers from scorched areas of the Shroud are entirely different from image fibers.

13) **No image formed under the blood stains.** The blood was on the cloth before the image formed.

14) **The image-formation mechanism did not damage the blood.** The image formation process was sufficiently mild that it did not destroy or damage the blood.

15) **The only image color visible on the back side of the cloth is in the region of the hair.** The color density of the hair image is much lower on the back side of the cloth than on the front. Any image-formation hypothesis must explain how the hair image could penetrate the cloth while body image did not.

16) **There is no evidence for tissue breakdown (formation of liquid decomposition products of a body).** This suggests definite time limits for image formation and cloth-body contact. Some reports from forensic pathologists suggest an upper time limit of about 30 hours.

17) **Any radiation that is energetic enough to cause direct color formation on the entire outer surface of a fiber would have to penetrate the entire volume of a fiber.** All of the image color is on the outer surface of the linen fibers. Not all radiation is energetic enough to penetrate a fiber and directly to cause the chemical changes that produce color. Low energy radiation must heat the materials that produce color to a temperature where it will decompose within the time available for image formation. The bond that breaks at the highest rate in all saccharides is the C-OH bond on the hydroxymethyl functional groups. That bond has an energy on the order of 80 kcal/mole. That is close to the energy of a photon of green light (about 530 nanometers). Light or other radiation less energetic than that can not color the cloth except by heating it. Radiation hypotheses for image formation have to explain penetration, defect formation, and chemical reactions. They also have to explain the discontinuous distribution of image color

The requirements make it apparent that no single, simple hypothesis will be adequate to explain all of the observations made on the Shroud. The impurity/Maillard hypothesis is proposed in an attempt to incorporate more observations into a single, complex hypothesis for image formation. It is important to recognize that Maillard colors *will form* every time amines and simple starches and/or sugars come together.

CHAPTER XII: KINETIC THEORY OF GASES AND IMAGE RESOLUTION

Chapter XI presented a hypothesis for image-color formation that involved only natural products from a dead body and impurities on a cloth.

All colors are a result of interactions between materials and visible light. The most common colored materials are chemical compounds that absorb and reflect specific parts of the spectrum; e.g., if they absorb blue and green light and reflect red light, you see a red color. A generalization can be made as follows: *colors are caused by specific kinds of chemical structures.*

The production of color from colorless materials requires chemical reactions. Some materials had to be available at the time of image formation that could react to produce the color. As discussed in Chapter XI, a dead body produces decomposition products, and ancient linen was produced in such a way that it contained specific types of impurities. The color-producing reactions of these types of compounds have been studied for many years. These materials can and do react to produce color.

In order to react, the volatile products from the body must diffuse to the cloth and react with the impurities. Rates of the chemical reactions depend on temperature, physical state of the reactants, and reactant concentration. A summary of gaseous-diffusion facts is presented below. Rates of diffusion depend on temperature, pressure, the mass of the molecules involved, and physical context.

Gaseous diffusion has been studied since the 18th Century. Generalizations have been made and used for predictions. Hypotheses for the observations have been tested extensively. The observations have been coordinated with thermodynamics, quantum mechanics, wave mechanics, and statistical mechanics. *Ab initio* calculations can now be made. The fundamental bases for observations on diffusion have now been advanced to the status of scientific laws.

Many comments have been made about gas-diffusion effects with regard to the image. Part of the problem involves semantics: in popular discourse, "diffuse" means "widely spread or scattered, dispersed." When a scientist discusses diffusion, he/she is thinking of specific laws of nature. The laws of diffusion should be used to test any hypotheses that involve the phenomenon. Much of the basic information can be found in any physical-chemistry textbook under "*Kinetic Theory of Gases.*"

Remember that the word "theory" is much different in science than popular discourse. A scientific theory is a hypothesis that has withstood repeated testing. The Kinetic Theory of Gases is an excellent example. Graham's *Law* of diffusion follows directly from the fundamental theory. No valid claims can be made about diffusion and image resolution without reference to the fundamental facts.

When the members of STURP first observed the image in 1978, it was clear that all of the color was on the tops of the highest parts of the weave: it did not penetrate significantly into the yarn (Figure III-1). We said that it did not show the effects of gaseous or liquid diffusion. If it had, the color density should have decreased exponentially with distance into the yarn. It did not: the color appeared to stop abruptly within a very few fibers near the top. This statement left the impression that gaseous diffusion could not be involved in image formation. Far from it, because some other phenomena can affect the distribution of reactants on a cloth. With regard to image resolution, the same fundamental theory applies to lateral diffusion and its effects on resolution as applies to penetration into a surface. The details of the theory are important.

Kinetic Theory of Gases:
The main components of the Kinetic Theory of Gases are the following:
1) Gases are composed of discrete molecules.
2) The molecules are in constant, chaotic motion.
3) "Pressure" is caused by the molecules hitting the walls of a container.
4) Collisions between unreactive molecules tend to be elastic.
5) The quantity we call "temperature" is a measure of the *average* kinetic energy of all of the molecules in a system.

If you put a molecule of mass "m" into a cubical box with a side length of "l," it will bounce back and forth among the walls. When it hits a wall at a velocity "u," its momentum will be mu, and its momentum on rebound will be –mu (total change in momentum is 2mu). In order to make another bounce against the same wall, it has to travel a distance of 2l, but it can collide with any of the walls. Its velocity has components along all of the x, y, and z coordinates. The change in momentum for one molecule at one wall (e.g., the x direction) per second will be,

$$\left(2mu_x\right)\frac{u_x}{2l} = \frac{mu_x^2}{l}$$

Considering all of the directions, the total change in momentum per molecule per second will be

$$\frac{2m}{l}u^2$$

The result allows predictions of observations from the fundamental properties of molecules. Expansions of the theory enable the calculation of molecular velocities, pressure, viscosity, mean free paths between collisions, and kinetic energies in the different modes (e.g., linear motion, vibration, and rotation).

In thermodynamics, the ideal gas equation of state for one mole of gas is $PV = RT$ where P is the pressure, V is the volume, R is the "gas constant", and T is the absolute temperature (in degrees Kelvin). Kinetic theory shows that $PV = \frac{1}{3}Nmc^2$ where c^2 is the

112

mean square of the velocities of the molecules at the experimental temperature. This leads to the fact that the total energy of a mole of gas, $E_{total} = \frac{3}{2}RT$.

According to the *principle of equipartition of energy*, each form of energy in a polyatomic molecule contributes $\frac{1}{2}RT$ to the average energy of one mole of gas.

The simplified calculations are good, but they are not perfect. In the real world some molecules are rotating but not vibrating while others do both. Accurate calculations required quantum theory, wave mechanics, and statistics.

Wave mechanics showed that, even at absolute zero, a molecule will still possess some energy, mostly vibrational energy. This energy is called the ***zero-point energy***[2] of the molecule. Here is where statistics enters. The ***most probable*** energy distribution among a collection of molecules will be

$$N_i = N_0 e^{-\varepsilon_i / kT}$$

where N_0 is the number of molecules in the lowest-energy state, N_i is the number in the ε_i energy state, and k is the **Boltzmann constant** (the molar gas constant, R, divided by Avogadro's number, the number of molecules in a mole).

So what does this have to do with image resolution? Many claims have been made that no diffusion process could yield the resolution observed in the image. Such claims should not be made or accepted without some realistic calculations and confirmatory experiments. Experiments can be designed according to known laws such that variables are separated for individual study. Some experimental results will be presented below.

Diffusion does not mean the same thing as "diffuse." The ***mean free path*** of a molecule will determine its diffusion properties. The mean free path is the distance the molecule will go between collisions with other molecules. Such calculations show that molecules can not go very far unimpeded at normal temperatures and pressures. The potential for image resolution is quite good.

It can be shown by a fairly simple argument that a molecule in a gas with n molecules per cm^3 will collide with $\sqrt{2}\pi v\sigma^2 n$ molecules per second, where σ is the molecular diameter and v is the velocity of the specific molecule. Notice that the diameter is ***squared***: it will be very important. When the size doubles, the number of collisions will go up by a factor of four. A large molecule will suffer many collisions. Cadaverine is many times larger than ammonia.

The numerical values needed for calculations/predictions are usually obtained from gas viscosity measurements, because (according to kinetic theory) the viscosity coefficient, η, is given by

$$\eta = \frac{1}{3}vld$$

where d is the density and l is the mean free path.

The greater the mean free path the more "diffuse" will be gas dispersion. ***But each gas will have its own mean free path under the same conditions of temperature and pressure.*** It is not valid to "assume" a large dispersion for a gas. Depending on its chemical and physical properties, it may not disperse very far.

As one quantitative example of diffusion, the ammonia molecule has a diameter of 0.297 nanometers. Its mean free path at 20°C and 750 mm pressure is 66 nanometers. That is only **0.0000066 cm** (0.0000026 inch). That is a very, very short distance, but ammonia is the lightest of the amines. All other amines will show less diffusion. Ammonia is only about 13% as heavy as an average body-decomposition amine. The most important decomposition products will have much better resolution potential.

Graham's Law of Diffusion follows directly from the Kinetic Theory of Gases, and it has the following form:

$$\frac{v_1}{v_2} = \frac{\sqrt{d_2}}{\sqrt{d_1}} = \frac{\sqrt{mw_2}}{\sqrt{mw_1}}$$

where v = velocity of diffusion, d is the density of the gas, and mw is the molecular weight of the gas. The velocity of diffusion of cadaverine (1,5-pentanediamine, one of the amines you smell over a dead body) is only 37% of the velocity of ammonia.

When gases are diffusing into one another, the concentration at any point (c) depends on the starting concentration (Q), time (t), and the ***coefficient of diffusion,*** Δ (also called diffusivity}.

$$\frac{dQ}{dt} = -\Delta\left(\frac{dc}{dx}\right)dydz$$

The diffusivity of a gas goes down as its molecular weight goes up; for example, the diffusivity of hydrogen (mw = 2) is 0.634 while that of oxygen (mw = 32) is 0.178.

The ***concentration gradient***, $\left(dc/dx\right)$, is the rate of change in concentration with the distance (x) This is the factor that causes a diminution in reactive-gas concentration as body-cloth separation increases.[2] Also, if there can be a large change in concentration in a small distance, resolution can be good. It is not valid to assume a diffuse image when gaseous diffusion is involved in the image-formation mechanism.

114

Heavy decomposition amines will have very steep concentration gradients as they diffuse into air. Their concentration in air will be greatly reduced as distance from the body increases. The diffusion-limited concentration gradient will be a major factor, but **not** the only one, in calculating the maximum resolution of an image-formation mechanism that involves diffusion.

Concentration gradients are also formed by depletion of reactant(s) as a result of the chemical reactions. Other mechanisms leading to loss of reactant and production of a concentration gradient include adsorption on the surface without reaction, diffusion through pores in the cloth, and non-color-producing parallel reactions.

Fibrous or particulate barriers can inhibit diffusion rates. Porous barriers increase the concentration of heavy reactive gases on the up-stream side of the barrier. This is the principle used in uranium-isotope separation by gaseous diffusion, and it would affect concentration gradients during diffusion through hair.

Increased concentrations result in increased reaction rates and loss through the pores of the cloth. Notice that the face image is surrounded by hair, a barrier, and it is the darkest part of the image [Cunico A., "Ricostruzione dell'immagine corporea dell'Uomo della Sindone: analisi preliminare della correlazione tra luminanza e distanza corpo-telo", Degree thesis, tutor G. Fanti, Dipartimento di Ingegneria Meccanica, Università di Padova, academic year 1999/2000]. Also, the hair appears to be the only easily visible part of the image on the back of the cloth, just as would be predicted for a diffusion barrier.

The important point to recognize is that blanket, qualitative statements about diffusion and resolution can **not** be supported by simple assumptions.

Decomposition of a human body:
Calculations predict that an image-formation mechanism that involves gaseous diffusion should show higher resolution as the body cools and heavier decomposition products begin to predominate. Some facts about body decomposition become important.

Different organs in the body produce decomposition products at different rates, and body temperature determines the times until a specific product appears.

A very large amount of quantitative information has been obtained by experiments on actual cadavers at the University of Tennessee. A recent paper by Vass et al. ("Decomposition Chemistry of Human Remains: A New Methodology for Determining the Postmortem Interval," J. Forensic Sci. 2002;47(3):542-553) presented a method for predicting the trade-off between temperature and time in building up concentrations of decomposition products.

The authors suggest using a value, "Cumulative Degree Hours (CDH)," together with analyses of decomposition biomarkers to calculate the time since death. You can immediately see how this is related to observations on the Shroud. We can estimate the times required at different temperatures for specific products to appear over the body.

The CDH system cumulatively adds the average temperatures for each 12-hour period. The lower the CDH for product appearance the more rapid is decomposition of the specific organ. Incidentally, the skin is the body's largest organ.

As one example, the CDH for a 10°C, constant-temperature tomb would be 20 per day. Their analyses showed that the decomposition products of muscle tissue increased until about 800 CDH, i.e., something less than 40 calendar days.

The authors state that "Human decomposition begins approximately 4 min after death." Some products appear very quickly, but each organ is different, enabling the medical investigators to calculate the postmortem interval (PMI) from analyses of the organs. The concentration of volatile decomposition products over a body increases with time and temperature; however, a body will normally cool fairly rapidly.

The authors note that: "Initial surveys indicated that the common, odoriferous amine indicators of decomposition, cadaverine and putrescine, would be useful biomarkers... While the concentrations of these compounds were quite abundant (>3 micrograms per mg of tissue) in some instances, the values (between corpses) were quite inconsistent." Although the heavy amines are not diagnostic for PMI estimations, the amounts and rates are important for considerations of image formation. The analyses give us some indication of available reactant(s). Tenths of pounds can be produced by a body, and the total will be delivered over a significant length of time. The latter point is important in a diffusion, image-resolution context.

Vass et al. also report that putrefaction (structural degradation) generally starts between 36 and 72 hours after death, although rates do depend somewhat on clothing. The Shroud shows no obvious signs of putrefaction products. Perhaps we can assume that the cloth could not have been in contact with the body for more than about three days.

Ammonia appears very quickly from lung tissue, and it disappears fairly quickly; however, it can diffuse over a large area during its time of production. Ammonia diffuses much more rapidly through air than do the heavier products; however, production of heavier decomposition products continues for quite a long time. Image formation could be a slow process.

Resolution experiments:
Maillard reactions were discussed in Chapter XI. The important point is that all of the reducing saccharides react rapidly with amines, including the decomposition amines from a body. This is a very definite phenomenon: given a reducing saccharide and a decomposition amine, they *will* react.

Reducing saccharides have been detected on the Shroud, and Pliny the Elder discussed the use of starch in the production of linen. When such a cloth is placed near a decomposing body, the amine/sugar *reactions will take place*. The final products are yellow-brown in color.

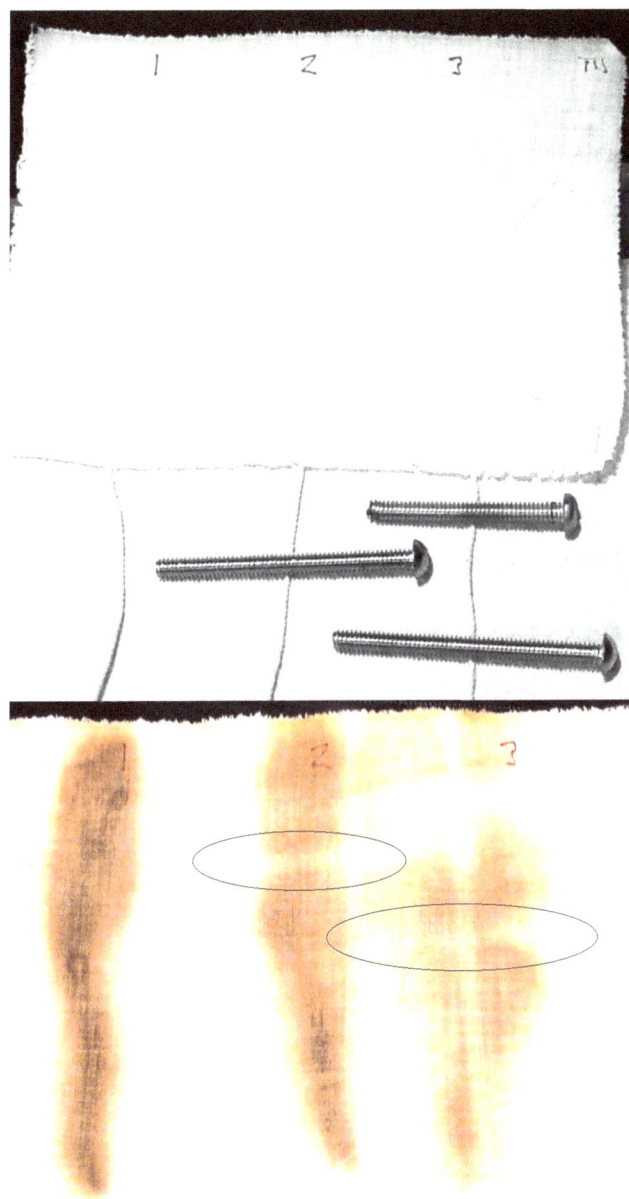

Figure XII-1: Setup and final image from an experimental test of resolution attainable by diffusion.

As discussed in Chapter XI, the chemical processes involved in color production between decomposition amines and saccharides are called Maillard reactions. A demonstration of some of the principles involved in diffusion and Maillard reactions is shown in figure XII-1.

The cloth (an old, often-washed, commercial muslin, i.e., cellulose) is shown at the top, marked 1, 2, and 3. It was saturated with a natural mixture of reducing saccharides and was then washed with *Saponaria officinalis* suds. Diethylenetriamine, DETA, mw 103.2, was used to simulate cadaverine, mw 102.2, because cadaverine is too foul to use. The amine was applied by saturating the cords with a 1:1 mixture of amine and acetone and allowing the acetone to evaporate. Bolts with 10-mm-diameter heads and 1-mm thread spacings were used to test resolution. The upper part of the photograph shows the physical arrangement of the bolts before the test.

The treated cloth was suspended over the bolts such that the left side touched the lower surface while the right side had a 12-mm clearance. This represents a significant difference in terms of diffusion paths.

The assembly was left in still room air at 22-23°C for 12 hours. No convection cells could develop with the top and bottom temperatures the same. Material transport would be by diffusion alone. A very faint yellow developed on the surface of the cloth; however, it rapidly developed the color shown as it was heated to 65°C for 30 minutes.

Figure XII-1 shows that the DETA had diffused from the cords through the vapor phase to react with the reducing saccharides on the treated cloth. The initial reaction was rapid; however, it did not produce colored products.

The images of the two lower bolts are indicated by the ellipses. The upper ellipse shows that the shaft of the bolt cast a sharp image. Some reduction in resolution can be seen between the left end of the ellipse at about 5-mm clearance and the right end where source to cloth clearance was about 8-mm. The lower ellipse shows a very definite and obvious reduction in resolution between a clearance of about 8 mm and 12 mm. It is not clear whether the thread pitch is visible on the cloth, because the weave has nearly the same dimensions; however, resolution is on the order of one millimeter.

These results can be compared with those shown in figure XI-1. The "hand" used in that test had been filled with rice, and it was heated to 43°C before the amine was painted on and the cloth was placed over it. Resolution was degraded by convection during the time the hand cooled to room temperature.

Some specific observations are the following:
1) The locations of the bolts can be seen in the diffusion pattern. The head of the middle bolt is clearly visible as a white area.
2) Striations are visible in the muslin: the weave is not homogeneous (just like the Shroud). This fact is emphasized in the developed "image." Notice the light band to the left of the 3 (3 was the sample with the greatest cord to cloth separation and consequently the widest diffusion pattern). Notice the much finer horizontal and vertical striations in the darkened areas. They look much like the weave pattern in image areas.
3) The higher the cloth above the amine source the more diffuse the shading. This illustrates the effect of diffusion.

118

4) The heating was sufficient to color some apparently clear areas where no significant amount of amine was expected to react. This suggests that much of the background color on the Shroud could be a result of coloration in the same surface impurity that enabled image formation. The Shroud shows "ghosts" in non-image areas. The lighter color of the back of the Shroud is probably a result of impurity separation, front to back, during the drying of the new piece of linen (see Pliny the Elder).

5) Color density was higher on the surface where evaporation was most rapid, i.e., where the saccharides concentrated. All of the fibers have a very similar color density, similar to the "half-tone effect" of the Shroud.

6) Color density varies with source-cloth clearance. Diffusion of the DETA into air produced a concentration gradient that affected the color density. Diffusion could explain the variations in color density of the image.

Slow production of decomposition products could be important for image resolution. We have all witnessed the phenomenon of "flooding." When rain falls too fast, water runs off of the soil. The same amount of rain on sand can soak in. The "permeability" of a material determines how easily fluids can penetrate it or flood it. When you put an impermeable plastic sack over your head, you will die within some minutes. When you put your head under a sheet, you do not die of asphyxiation: air diffuses through the pores of the cloth. If a large amount of amine were quickly released inside a cloth, it would flood the surface: a diffuse area of interaction would result. This may be visible where ammonia diffused across the surface of the Shroud. A body decomposing at a natural rate would not produce reactive gases at a rate that would flood a cloth surface.

Since I did not have access to a fresh corpse, I had to use saturated sources for my experiments.

- - - - -

[1] Convection can "stir" the gases and increase diffusion rates; however, convection is driven by temperature differences. The density of a gas decreases as it warms (the reason hot-air balloons rise). As warm air rises, cooler air flows downward to replace it, creating "convection cells." A body cools fairly quickly during the first few hours after death, before significant amounts of decomposition products have been formed. No significant amount of convection will occur when the body and cloth reach the same temperature.

[2] The zero-point energy is important in calculations involving the fractionation of the different-mass isotopes of the same element. Many illogical, poorly-supported claims have been made about ^{14}C enrichment as a result of the 1532 fire. The claims need careful study.

Maillard reactions:

Maillard reactions occur in three stages: 1) The condensation of compounds that contain amino groups with reducing sugars. The sugars lose water to form N-substituted aldosylamines (colorless). 2) The aldosylamines are unstable and undergo an "Amadori rearrangement" to form ketosamines. 3) The ketosamines undergo subsequent complex dehydration, fission, and polymerization reactions. These products contain many conjugated carbon-carbon double bonds, and they are colored. A well-known example is the brown crust on bread.

The final Maillard reaction products are very close to those of caramelization; indeed, one of the Maillard paths is a simple caramelization reaction that is catalyzed by amino groups. However, the Maillard reactions can take place at much lower temperatures and/or shorter times.

CHAPTER XIII: CONSERVATION

The Shroud of Turin is a "chemical object." Every part of what we see and can feel is made of chemical compounds, and all of those compounds have been studied extensively. The main component of the Shroud is cellulose. Of all of the organic compounds that occur in nature, including the building blocks of all of the animals, fish, and us, cellulose is by far the most abundant. It should not come as any surprise that its properties are well known among chemists.

Cellulose is a member of a large group of chemical compounds known as "carbohydrates." Cellulose appears in a very pure form in cotton, but all carbohydrates are complex assemblages of sugar units of different kinds. It is estimated that about 1.1 *trillion* kilograms of cellulose exists at any given time as a result of photosynthesis by plants. There is a field of chemistry entirely devoted to the study of carbohydrates.

Cellulose is made of long chains of the sugar d-glucose (dextrose) that are linked together glycosidically, i.e., through oxygen links. The glycosidic linkages of cellulose can be broken by several well-known processes. As linkages are broken, the fibers weaken. Some microorganisms ferment cellulose and destroy its structure. Aerobic organisms produce carbon dioxide from the cellulose; anaerobic organisms in the mud at the bottoms of swamps produce methane. If it were not for these organisms, we would be buried in cellulose, but cellulose is not easy to decompose by fermentation, a fact that any expert on sewage disposal can confirm. If kept dry and out of direct light, cellulose lasts indefinitely. You can still read magazines from old garbage dumps.

A panic about the stability of the Shroud caused a poorly conceived and executed "restoration" during June and July of 2002. Giuseppe Ghiberti wrote the first publication on the "restoration" in a booklet of photographs, *Sindone le immagini 2002.* He attributed the motivation for the hurried cleaning, scraping, and vacuuming to Professor Alan Adler. Ghiberti stated that: "Adler worried about the possibility of the damaging effects of material from the 1532 fire that was trapped under the patches applied by the nuns of Chambéry. When he spoke during the Committee meetings, he did not retreat from the most advanced hypotheses. One of these (certainly not new, as mentioned above, but backed now by the weight of his authority) was to remove the patches and the Holland cloth." Unfortunately, Al died 12 June 2000, and he could not maintain control over the juggernaut he had put into motion. He was the only chemist on the committee, and the custodians of the Shroud carefully kept their plans secret from everybody. This illustrates the dangers of appeals to authority.

Ghiberti also reported that Dr. Carla Enrica Spantigati said: "The nearly five hundred years spent by the Shroud in the company of its Holland cloth and its patches had established a characteristic of a stable tradition in the life of the Shroud ... it would be wise to continue the present situation." This was excellent advice, but the committee did not question their own knowledge of the "situation." Adler had so convinced the committee that dire things would happen to the Shroud as a result of "the autocatalytic decomposition of cellulose" that their panic overwhelmed the words of caution from Dr. Spantigati.

No expert on carbohydrate chemistry or the chemical problems associated with the conservation of cellulosic materials was consulted. The committee did not understand science, and they made no attempt to "assemble all data." This led to a greater disaster than the fire of 1532.

AUTOCATALYSIS:

The fact that the Shroud survived the fire of 1532 by 470 years without obvious acceleration of any degradation processes should provide all of the proof necessary that autocatalysis is not a significant problem. None of the products observed during the detailed chemical studies made by STURP in 1978 could be involved in autocatalysis, and they should cause no fear for the longevity of the cloth. It is depressing that the Conservation Committee did not study the literature on the Shroud before taking irreversible action. The chemical investigations of 1978 had been published.

Based on the facts of chemistry, the Shroud of Turin is not now and has not been for the past 470 years in danger of catastrophic autocatalytic decomposition.

The Chemistry of Autocatalysis:
Autocatalytic chemical reactions are those in which the rate increases as the amounts of reactants decrease, i.e., while the materials are reacting. The most important single factor in predicting effects is the *temperature*. When the temperature changes, the rate changes. The only severe heating episode the Shroud has suffered was during the fire of 1532. Any autocatalytic decomposition that occurred then has long since stopped as the Shroud is stored at normal temperatures.

Autocatalysis can not rigorously be discussed without some reference to basic chemical kinetics and mathematics. The simplest form of the fundamental equation that describes an autocatalytic process is the following:

$$\frac{d\alpha}{dt} = k\alpha^p (1-\alpha)^q Z e^{-\frac{E}{RT}}$$

where α is the fraction reacted at any specific time, t. The derivative, $d\alpha/dt$, is the rate of the reaction. E is the "Arrhenius activation energy," and Z is the "Arrhenius pre-exponential." Each applies **only** to a single specific, consistent reaction. The value of the "rate constant," k is different at each specific temperature: It is a *constant* only at one temperature, and it applies **only** to one specific reaction. The values of E and Z are determined from a large number of *k* measurements at different temperatures. Predictions of the Shroud's expected lifetime can not be made on the basis of a single rate constant. ***Observations made during a scorching event can not be applied to rates at normal temperatures.***

122

E, Z, and k are the most important values in a discussion of rates and associated lifetimes of materials. All of these values have fundamental meaning in the chemical reaction. R is the "gas constant (1.9872)," a universal constant that applies to many disparate physical and chemical processes, and it is known with great accuracy and precision. T is the *absolute* temperature, expressed in degrees Kelvin (0K = -273.2°C). The exponents p and q allow the prediction of the position of the maximum rate in an autocatalytic process, i.e., the amount reacted at the maximum rate - *at constant temperature*. Exponents higher than 2 are extremely rare.

Examples of simple and autocatalytic rate curves are shown in figure XIII-1. Notice that the rate increases with time in the autocatalytic curve, **at constant temperature**, until it reaches a maximum reaction rate. Then the rate decreases. However, **the initial rate at any temperature is much lower than the maximum rate.** The chemical decomposition rate of cellulose is essentially zero at room temperature.

When cellulose is decomposing autocatalytically at high temperature, the rate can be returned to its initial value by cooling. The safest way to preserve cellulose is by keeping it cool but not too cold.

FIRST-ORDER RATE CURVE   AUTOCATALYTIC RATE CURVE

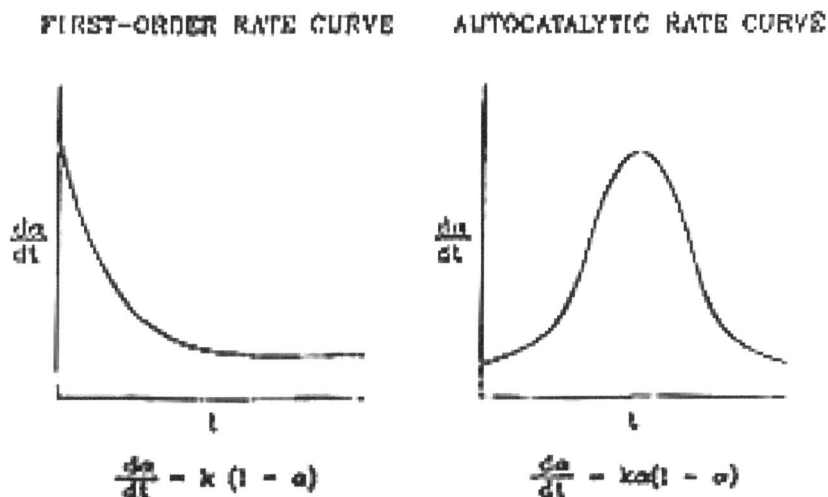

$$\frac{d\alpha}{dt} = k(1-\alpha)$$

$$\frac{d\alpha}{dt} = k\alpha(1-\alpha)$$

Figure XIII-1: Ideal rate curves for normal (left) and autocatalytic (right) chemical reactions.

The activation energy, E, is closely related to the strength of the reacting chemical bond. Strong bonds tend to show high activation energies when they react; weaker bonds show lower activation energies. Catalysts increase rates by lowering the apparent activation energy of a reaction.

The bond that breaks at the highest rate in the initial pyrolysis of cellulose is the C-OH bond on the hydroxymethyl functional group of the glucose units that make up the

123

cellulose polymeric chains. That bond has an energy on the order of 80 kcal/mole. It is not a weak bond.

Activation energies in solids, *especially crystalline solids like cellulose,* are higher than the values for the same material in a solution or melt, because a crystalline lattice is stabilized by its ordered structure. The rate constants for cellulose decompositions predict extremely slow rates, so slow you can call them zero.

The major cause for autocatalysis in solid cellulose decomposition is the destruction of crystalline order when the material is heated to a high temperature. The difference between E in a melt and in the crystalline solid is often close to the latent heat of fusion. ***Melted materials decompose more rapidly than the same material in a solid phase.*** When the material is cooled below the melting point, autocatalysis stops. Rates in the normal cellulose solid phase are essentially zero in the absence of catalysts, short-wavelength light, or some microorganisms. Chemical autocatalysis is discussed below.

The initial decomposition rate of a crystalline solid depends on crystal perfection. When crystals are put under stress, they develop high-free-energy defects, and decomposition is much faster at the defects than it is in the parent material. If autocatalysis were a real problem for the Shroud, significant differences should have been observed in damage around the stressed and strained stitching of the patches. I looked for such evidence in 1978. There was no sign of accelerated decomposition, indeed there was no sign of any autocatalysis. Autocatalysis is not a real hazard for the Shroud. The Conservation Committee never contacted any of the chemists involved in the STURP studies after the death of Al Adler. They certainly should have done that.

The structure of pure, uncatalyzed, crystalline cellulose does not begin to dehydrate at a significant rate until about 300°C, and it is an extremely slow process at normal temperatures. Samples of wood and textiles exist from thousands of years ago.

Some reactions involve *chemical* autocatalytic rate processes in which a product (or products) of the primary reaction is a catalyst for the reaction: The rate increases as catalytic products accumulate ***at constant temperature.*** Small changes in the composition of the reacting material can have very large effects on the values of E and Z, as in any case of catalysis. This fact should be remembered in the discussion on the addition of thymol to the Shroud. If the process is cooled, the rate will be much lower, depending on phase changes, the magnitude of E, and retention of the catalytic product(s).

The possibility for chemical autocatalysis in linen depends on the products of cellulose decomposition. Feigl and Anger described the effects of heating cellulose as follows: "When cellulose is heated it decomposes and the resulting superheated steam reacts with unchanged cellulose to produce hexoses, which in turn hydrolyze to give hydroxymethylfurfural." The only important chemical catalyst for the autocatalytic degradation of cellulose at high temperatures is ***superheated steam.***

124

If catalytic products are gases, they can escape from the reacting zone after cooling. Superheated steam does not exist at normal temperatures, and the only important autocatalytic process in cellulose stops until another heating episode. There is no "memory effect." The Shroud should be as stable at room temperature as any other sample of linen.

Although the primary decomposition process of all carbohydrates, including cellulose (linen), is dehydration, linen adsorbs and absorbs water from the atmosphere, depending on the relative humidity and temperature. When linen is heated, water immediately begins to be desorbed, the linen dries out. This can have a major effect on measurements of the strength and weights of large pieces of cloth. It is difficult to determine the difference between adsorbed water and water of reaction. It may require measurements involving kinetic isotope effects with deuterium. I do not believe it would be logical to attempt to monitor the condition of the Shroud by observing the rate of appearance of water in its containment vessel.

Many chemical processes involve "equilibrium": The reactions can proceed in both the forward and reverse directions. When one product can be eliminated (e.g., pumped out as a gas or absorbed by a desiccant), the reaction proceeds in that direction. The Shroud should not be kept too dry, or dehydration will be promoted. Even then, the rate would be quite low. The major danger in desiccation is producing a brittle cloth.

More detailed studies have shown that the major secondary products of the thermal decomposition of cellulose are formaldehyde, furfural (2-furaldehyde), hydroxymethylfurfural (5-hydroxymethyl-2-furaldehyde), carbon monoxide, levulinic acid (4-oxopentanoic acid), and 3-pentenoic-γ-anhydride. None of these can be expected to be a significant catalyst for the decomposition of linen. Indeed, formaldehyde, furfural, and hydroxymethylfurfural are reducing agents, antioxidants. Furfural inhibits the growth of molds and yeasts. Scorched areas are less likely to show microbiological attack.

I found that cellulose produced relatively more levulinic acid than furfural when compared with pentoses. That is not a problem in the context of the Shroud. Levulinic acid is a solid at room temperature (mp 37.2°C); it partially decomposes at its boiling point (245°C), it decomposes much faster at higher temperatures and in oxygen, as during the fire of 1532; and it is an extremely weak organic acid that would have little effect on cellulose. It has a significant vapor pressure, and it gradually vaporizes from any surface.

I measured activation energies and decomposition mechanisms for the individual decomposition products of several carbohydrates. One example, the rapid, *catalyzed* decomposition of the pure hexose sugar levulose, is shown in the figure XIII-2. The sugar units in linen are glucose, a similar hexose.

I caught the products of decomposition on a thin-layer chromatographic plate as a function of the temperature at which the levulose was decomposing. I then "developed" the plate by allowing 1,2-dichloroethane to migrate up it from bottom to top. The more soluble and/or less strongly adsorbed products were pushed farther up the plate by the solvent. R_f is the ratio between the distance the single product migrated and the distance to the leading edge of the solvent. The distances the products migrated and chemical spot tests made it possible to identify them.

This reaction has been chosen to illustrate a ***catalyzed*** carbohydrate decomposition. It would be hard to find a more rapid decomposition of the kinds of molecules seen in cellulose. It was catalyzed with concentrated phosphoric acid, a nonvolatile, strong, inorganic acid. The vertical axis shows the R_f (identity) of each specific product; the horizontal axis shows the temperature at which the product appeared. II is levulinic acid, which decomposes to produce VI, 3-pentenoic-γ-anhydride. The important observation is that ***no*** condensable products

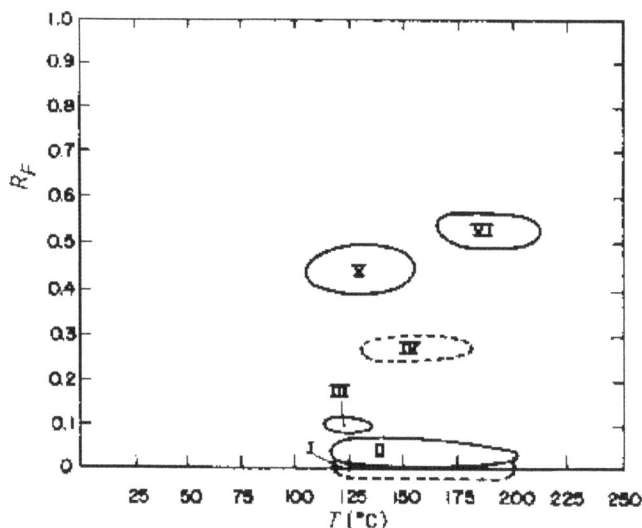

Figure XIII-2: Pyrolysis-Thin-Layer-Chromatography plate for the strongly catalyzed decomposition of levulose.

appear below about 115°C, ***even when the reaction is strongly catalyzed.*** This kind of catalyzed reaction is similar to that responsible for the destruction of books that are made with cheap, acid paper. Claiming analogy between such books and the Shroud is mischievous.

Adler wrote the following comment in an attempt to direct attention toward conservation: "Previous chemical reactions on the cloth, e.g., the retting process in manufacture of the linen, the known historic fire and its extinguishment, and previous display and storage procedures, have left a variety of chemical structures on the surface that can act as oxidants and also as catalysts. For example, the acidic structures produced by previous oxidative activity can strongly promote various types of autocatalysis" It certainly got attention, but Al could not direct that attention into productive avenues after his death. This is the kind of statement that initiated the hurried, secretive, disastrous "restoration" of the Shroud in 2002. It obfuscates the fact that the Shroud might have survived about 20 centuries of storage under uncontrolled conditions. It certainly survived devotion for 470 years.

126

I have not been able to observe any chemical autocatalysis in cellulose at normal temperatures. Observations and descriptions of the Shroud through the 470 years since the fire of 1532 do not support fear of catastrophic decomposition of the cloth. When I was taking tape samples as part of STURP in 1978, there was no evidence for weakening of the cloth: It was as strong as the Holland cloth and the patches. The Shroud is still in its "induction time," (a reaction rate very close to zero) and it is in no danger of immediate degradation. When I say "close to zero" I am speaking as a chemist. All chemical rates are modeled with the Arrhenius equation, which does *not* allow for a zero rate except at absolute zero temperature. It would probably be better or less confusing just to say "zero."

If Shroud deterioration is still a worry, one practical way to slow the rate is to keep the Shroud cold. That also has the advantage of reducing microbiological attack. As in the case of the use of "inert" atmospheres, storage at reduced temperature should carefully be considered. Too low a temperature could cause physical stress and might cause fibers to fracture. It would probably cause the thin coating of image color on the fibers to be loosened in some areas.

As a rule of thumb according to the Arrhenius expression, rates of normal reactions are increased by a factor between two and three for each 10°C increase in temperature. The effect is much larger when a phase change is involved, e.g., melting. Some moderate cooling could have a significant effect on prolonging the life of the Shroud. Severe freezing could damage the cloth and image.

Effects of the "Restoration" on Future Chemical Studies:
During the 1977 United States Conference of Research on The Shroud of Turin, I presented information on how existing photographs of the damage from the fire of AD 1532 gave us a large amount of information on the chemical composition of the cloth and image. The information in 1977 made me doubt that the image was painted, and that made the project interesting; however, the "restorers" of 2002 completely ignored the valuable chemical information contained in the areas they changed.

If the image had been painted or retouched, some foreign materials had to be added to the cloth. The pigments and vehicles (e.g., glair, gums, and glues) would have been subjected to a violent "chemical test" during the fire. The temperatures, temperature gradients, pyrolysis products, and water used to extinguish the fire would have changed the chemical composition of most foreign materials. Before going to Turin in 1978, we did many experiments on the stability of the painting materials. Records of these tests and the materials used have been published and archived, and they illustrate the damage done by the restoration.

Although the fire of 1532 nearly destroyed the Shroud, it created opportunities for many types of chemical studies. We would never use the same destructive methods of observation on an undamaged relic, but misadventure handed us a goldmine of information in 1532. The important fact is that, before the restoration, we could look at the chemistry of specific locations on the Shroud where scorches intersected image, blood, serum, and water stains. The restoration totally destroyed any chemical information at those intersections. All of the material was scraped away.

A large body of chemical information is available on the interactions among reactive pyrolysis products and known or suspected Shroud components. For example, McCrone claimed until his death that the image was painted with hematite. Experiments we did before 1978 proved that the pyrolysis products from the fire would reduce red hematite to black magnetite on the Shroud. No such effect could be observed on the Shroud during the hurried observations of 1978, but I took tape samples from all important areas for later study. Unfortunately, the tape samples were inadvertently returned to Turin after the death of Al Adler, and they have been lost to scientific observation. Authorities in Turin refuse to discuss either the tape samples or any chemical data they might have collected during the "restoration" of 2002.

All paints that were used during or before medieval times are changed by heat and/or the chemically reducing and reactive pyrolysis products of the cloth (e.g., formaldehyde, furfural, organic acids, CO, etc.). Some medieval painting materials become water soluble, and they would have moved with the water that diffused through parts of the cloth at the time of the fire. A huge amount of chemical information existed in the scorches.

Most organic colors are much less stable than cellulose (linen) and the normal inorganic pigments. Experiments in 1977 showed that scorch lines in impurities precede the scorches in pure linen. Most organic materials, including natural products, change in predictable ways in response to heating and the known products of cellulose pyrolysis. We even tested squid ink, which had been reported being used in ancient times. The products of the reactions can be extracted from cloth and used to prove original compositions. Such information was important for suggesting the chemical composition of the image. Most possibilities for studying the effects of the fire on image materials were destroyed by the restoration of 2002.

Visual observations of the Shroud in 1978 indicated that image color did not move with the water. It could clearly be seen that blood had been denatured and colored at intersections with the scorches. The option for sampling those areas as a function of temperature has been lost.

Some unidentified products did move with the water. We had counted on the tape samples and possible future direct studies on the scorch/water-stain areas of the Shroud for detailed chemical confirmation of what did and did not move with the water. Aldo Guerreschi has proposed that some of the water stains on the cloth had formed before the fire of 1532. Chemical studies on the areas could have given significant historical information. Water percolating through new scorches would carry the water-soluble pyrolysis products with it. Water percolating through the cloth earlier in its history would have carried water-soluble impurities from the cloth's production. It may now be impossible to make appropriate observations on the water stains, and the authorities in

128

Turin certainly do not appear to appreciate what historical information can be learned through chemical analysis. Cooperation from Turin has been abysmal.

Lignin is a structural polymer that is found in all plants, including flax. Linen is bleached in an effort to remove as much lignin as possible, but ancient bleaching techniques were not very successful. As discussed in Chapter VII, it is still quite easy to find some lignin in linen that was made after about AD 1200. The technology of the Shroud clearly puts it into the category of "ancient linen."

The rate data discussed in Chapter VII show that the lignin in the Shroud is significantly different from modern lignin: no vanillin can be detected in it. It shows very similar properties to the linen used to wrap the Dead Sea scrolls. This observation would suggest that the linen of the Shroud is very old, casting doubt on the accuracy of the 1988 date.

The accuracy of our chemical-kinetics measurements (Chapter VII) could have been tested with samples from the scorches. The obvious question is how temperature gradients affected the lignin at the time of the fire. Sensitive instrumental analyses applied to the thermal gradients at the leading edges of scorches could have given detailed information on the kinetics and mechanisms of lignin degradation on linen. The restoration eliminated or disturbed the sampling areas, and any study of linen/scorch intersections may now be inconclusive or impossible.

If we knew its real age and how much vanillin was left in the cloth, maybe we could estimate the temperatures where it had been stored. I have used the technique with explosives. Was it in Edessa in a wall that was often in the hot sun? It is too bad that the custodians of the Shroud do not feel the need of outside knowledge or consultation. They obviously have not given the Shroud much thought from a chemical perspective.

It is extremely unfortunate that all of the scorch areas that contained chemical information have been destroyed by the "restoration." Turin refuses to give us information on the way the removed material has been stored. It would be some comfort if scrapings were segregated with regard to location, but the effects of temperature gradients would still be lost.

One powerful chemical analytical method that has become more easily available since the early stages of Shroud science planning is Electron Spectroscopy for Chemical Analysis (ESCA). It enables the determination of the structures of organic materials on the top few molecular layers of a sample. Such observations were the "Holy Grail" for those of us who were interested in Shroud science, especially image-formation. We might have been able to identify the actual chemical processes that produced the color of the image. Since the surface of the entire Shroud has now been disturbed, such an approach is unlikely to succeed. This is a terrible, discouraging loss for Shroud chemists.

The persons involved in the "restoration" of June and July 2002 did not consult any chemists or chemically-oriented textile conservators; consequently, any future studies will be much more difficult and expensive as a result of the thoughtless operations, and a large amount of potentially critical information has been lost forever.

Microbiological Attack:

Garza-Valdes and Mattingly claimed that, without looking at the Shroud, they could swear that the entire surface was coated with organisms and a "bioplastic polymer." I believe that they misidentified the gum coating on the Raes, and presumably radiocarbon, samples as a bacterial polysaccharide. The observations by STURP in 1978 would have detected such materials, but no significant amounts were found.

The unfounded claims for microbiological attack may be responsible for another disaster. Although the authorities in Turin refuse to discuss the unfortunate radiocarbon age determination of 1988, there is significant evidence that the sample was taken from an anomalous location. As discussed in Chapter IX, there is a justifiable doubt that the published date corresponds to the date the cloth was produced. At some point, the authorities will recognize this fact. A well-controlled radiocarbon sampling operation and analysis will be needed to confirm or refute the 1988 result. It may be difficult or impossible to obtain.

Some pristine samples from the main part of the Shroud will be required. The 1988 tests should have used scorched material, and it is not clear who decided to cut cloth from an undamaged area. It was a big mistake. But another serious mistake was made in 1988. "Le Prelevement du 21-4-1988; Etudes du Tissue: Number 1 [in Actes du Symposium Scientifique sur le Linceul de Turin, Paris 7-8 September 1989] contains a detailed article by Giovanni Riggi on the sampling operation. One piece of information was disturbing.

Riggi's article explains Professor Franco Testore's "complex criteria" for choosing the sampling location. Fluorescence photographs of the area had been taken in 1978, and it was known to be anomalous (see figure IX-11). As if it were not bad enough that they ignored all previous studies and chose the worst location for sampling (leading to the mischievous date published in 1989), Riggi mentioned another procedure that was undertaken without any consultation with chemists.

The article states: "The presence of humidity, dust, fungi, etc., was combatted during the 21 April 1988 session by a highly complicated and ingenious application of thymol to the reliquary." The original article indicates that Riggi had found fragments of mites on the cloth, and he and Testore were worried about all of the possible kinds of attacks on the cloth - - - except chemical. The Commission on Conservation worried about nonexistent catalysts, but non-chemists felt free to put one into the reliquary in 1988. The cloth had survived 470 years of less than professional conservation, but it is not clear that it can survive such thoughtless "help." Incidentally, we could not see any microscopic effects of fungal attack in 1978.

Riggi, Testore, and Gonella (none of whom are chemists) decided to treat the reliquary with thymol to sterilize it. They worried about the effect of thymol on the "unknown" composition of the image, but they did not consider its effects on all other materials. They sterilized the reliquary before the Shroud was replaced in it. They give the times during which bags of thymol were in the reliquary, but there is no way to estimate how much vapor reacted or was adsorbed on the walls of the container. Thymol has a high vapor pressure, and it adsorbs strongly to any type of surface. The reliquary was lined with wood. Wood has a cellular structure and composition that favors absorption of materials like phenols. You can think of it as blotting paper for thymol.

Surface areas are routinely measured in science by adsorbing vapors on the materials and measuring amounts (usually called the BET method). A significant amount of thymol could have adsorbed on the wood, and wood has a large cellular surface area. More thymol would have reacted with the cellulose and more reactive hemicelluloses, lignin, and plant gums of the wood. No analyses were done. The amount of thymol left in the reliquary is unknown, but it is critical.

Some thymol would have desorbed and transferred onto the Shroud rather quickly. The desorption would reduce the free thymol concentration, reversing the equilibrium of thymol-wood reactions. This would provide more thymol for reactions with the Shroud. Given enough time, reactants keep reacting until they reach their lowest energy state or equilibrium. I can not estimate the composition of the Shroud as it was taken from the reliquary. Apparently the persons involved with the 1988 sampling fiasco did not try.

Thymol is a "phenolic" compound, closely related to carbolic acid (phenol). The use of thymol shows a complete irresponsible ignorance of chemistry. Many superbly qualified chemists live in Europe and the United States, and some of them have had years of experience with the Shroud. They *care* about the Shroud. Many specialists on carbohydrates would have been happy to consult free of charge. Why were none asked about the long-term effects of thymol on cellulose (linen)? On iron compounds?

Thymol is also called "thyme camphor." It is obtained by steam distillation of different species of plants of the genus Thymus or Ajowan. ***All of its carbon is modern carbon.*** Each modern thymol molecule that is grafted onto the linen will reduce the apparent age of the cloth by some amount. A significant amount of reaction will totally destroy the option for making accurate radiocarbon age determinations on the cloth. The report by Riggi does not allow any estimation of the amount of thymol that is now grafted to the Shroud. The thymol that has reacted with the linen can not be removed by any safe method.

One hope for a valid radiocarbon sample is the carbonized cloth that was scraped off during the "restoration." Elemental carbon is extremely stable to chemical attack, and it can be cleaned very thoroughly. Depending on how much is available and how well it has been segregated and preserved (top-secret Turin information), it could be used to obtain a valid date.

Being a "phenol," thymol has a -OH group attached to a benzene-type ring. The aromatic ring makes phenols more strongly acid than the alcohols. They are very polar, and that is why they adsorb so strongly and react irreversibly with so many materials. Think of the phenol ("carbolic acid") you use on cold sores. Think of what pure carbolic acid does to your tissues. What a lot of carbolic acid does to materials, a little thymol will do to the Shroud over centuries.

One reason thymol is effective as an antiseptic is that it reacts with amino groups. It can be expected to react with the proteins of the blood on the Shroud. Among other things, it will denature the blood, making any genetic tests or blood-typing impossible. These types of products also result in dyes. Much of the "coal-tar" dye industry is based on such reactions. It was just plain stupid to allow any thymol in the vicinity of the Shroud, which the custodians presumably wanted to preserve in its original appearance.

Phenols also form esters with organic acids. Al Adler said he found acid groups on the image. They would react with thymol.

Given enough time, phenols will form ethers with other -OH groups, and a major part of cellulose is hydroxyl groups. Cellulose is made of sugar units. These ethers are very stable (that is why they form, even under unfavorable conditions), and you can not reverse their formation without destroying the cellulose. Hydriodic acid (HI) is usually used to open phenolic ethers. HI would destroy the structure of the cloth.

The 1978 STURP x-ray fluorescence study showed a wide range of iron concentrations on the Shroud, up to almost 60 $\mu g/cm^2$. Many phenols give intense colors with iron compounds. Some phenolic compounds are used to detect microscopic amounts of iron compounds. Depending on the amount and specific phenolic compound and enough time, the whole cloth might turn blood red, blue, green, brown, or violet. It is ironic that the custodians encouraged the superficial destruction of the Shroud and elimination of centuries of history trying to remove traces of "catalysts" that did not even exist while allowing addition of traces of an active phenolic compound.

Thymol will react with the shroud! Was the Shroud lucky enough to avoid massive effects? Only time will tell. It would have been much better to think about the effects of thymol before it was used in any amount. The result of its use may be a cloth on which you can not get an accurate date or that colors so much you can not see the image. Future generations may honor the bright-red (well-preserved) Shroud, telling their children how it used to show the image of a crucified man. A few arrogant "experts" who are certain they know everything important can do great damage.

It would be extremely laborious but possible to separate the most carbonized fibers from the mass of scrapings under a microscope. Pasteur used a similar technique to separate optical isomers. Perhaps some of the "restorers" could be sentenced to this operation for a few years.

I can observe carbonized fibers on sampling tapes from 1978, and some appear to be completely carbonized. Theoretically, it might be possible to separate the carbon from organic detritus by flotation. The approach could be tested on a very small scale with gradient-density tubes. Systems have been published that use a thermal gradient in a tube filled with a heavy organic liquid. Such liquids would wet the carbon, reducing the number of adhering bubbles. The same sample can be used many times in different pure organic liquids. Unfortunately, such testing would require a few milligrams of sample and cooperation by the custodians in Turin. We have a saying in the propellants field: "With sufficient thrust, pigs fly just fine." With sufficient pressure, even Turin might cooperate.

If enough carbonized material can be separated from the mass of scrapings, it can be cleaned in concentrated nitric acid. The acid attacks all carbohydrates and phenols much more rapidly than elemental carbon.

Having seen the results of the 1988 sampling and the secret 2002 "restoration," I have little hope that the custodians of the Shroud in Turin will volunteer to help with any reasonable, rigorous scientific studies.

Shroud studies move very slowly. Perhaps in a century or two, after the persons responsible for the most recent fiascoes are gone, some persons who care enough to risk embarrassment will control Shroud studies and conservation. Some generation may yet be able to make detailed scientific observations. This generation is totally stopped by the siege mentality of the authorities in Turin.

GENERAL REFERENCES

No painting pigments or media scorched in image areas or was rendered water soluble at the time of the AD 1532 fire.

1) R. N. Rogers, "Chemical Considerations Concerning The Shroud of Turin," in Kenneth Stevenson (Ed.), *1977 United States Conference of Research on The Shroud of Turin,* 23-24 March 1977, Albuquerque, NM, USA, Holy Shroud Guild, 294 East 150 St., Bronx, N. Y. 10451 (pp. 131-135).

2) E. J. Jumper, A. D. Adler, J. P. Jackson, S. F. Pellicori, J. H. Heller, and J. R. Druzik, "A comprehensive examination of the various stains and images on the Shroud of Turin," *ACS Advances in Chemistry, Archaeological Chemistry* **III:205,** 447-476 (1984).

3) L. A. Schwalbe and R. N. Rogers, "Physics and Chemistry of the Shroud of Turin, a Summary of the 1978 Investigations," *Analytica Chimica Acta* 135 (1982), pp.3-49.

Direct microscopy showed that the image resided *only* on the topmost fibers at the highest parts of the weave.

S. F. Pellicori and M. S. Evans, "The Shroud of Turin Through the Microscope," *Archaeology*, January/February 35-43 (1981).

The color density of any specific image area depends on the batch of yarn that was used in its weave.

1) V. D. Miller and S. F. Pellicori, "Ultraviolet fluorescence photography of the Shroud of Turin," *Journal of Biological Photography* **49**, 71-85 (1981).

2) http://www.shroud.com/pdfs/rogers2.pdf.

Adhesive-tape samples.

1) W. C. McCrone and C. Skirius, "Light Microscopical Study of the Turin 'Shroud,' I," *Microscope* **28**, 105 (1980).

2) E. J. Jumper, A. D. Adler, J. P. Jackson, S. F. Pellicori, J. H. Heller, and J. R. Druzik, "A comprehensive examination of the various stains and images on the Shroud of Turin," *ACS Advances in Chemistry, Archaeological Chemistry* **III:205,** 447-476 (1984).

3) http://www.shroud.com/pdfs/rogers2.pdf
.

The color of image fibers appears only on their surfaces.

J. H. Heller and A. D. Adler, "A Chemical Investigation of the Shroud of Turin," *Canadian Society of Forensic Science Journal* **14** (1981), pp.81-103.

Instrumental and microchemical methods of analysis.

1) L. A. Schwalbe and R. N. Rogers, "Physics and Chemistry of the Shroud of Turin, a Summary of the 1978 Investigations," *Analytica Chimica Acta* 135 (1982), pp.3-49.

2) R. Gilbert Jr. and M Gilbert,. "Ultraviolet-visible reflectance and fluorescence spectra of the Shroud of Turin," *Applied Optics* **19**, 1930-1936 (1980).

3) S. F. Pellicori, "Spectral properties of the Shroud of Turin," *Applied Optics*, **19** (1980),

pp. 1913-1920.

4) J. S. Accetta and J. S. Baumgart, "Infrared reflectance spectroscopy and thermographic investigations of the Shroud of Turin," *Applied Optics* **19**, 1921-1929 (1980).

5) R. A. Morris, L. A. Schwalbe, and J. R. London, "X-Ray Fluorescence Investigation of the Shroud of Turin," X-Ray Spectrometry **9**, 40-47 (1980).

6) J. H. Heller and A. D. Adler, "A Chemical Investigation of the Shroud of Turin," *Canadian Society of Forensic Science Journal* **14** (1981), pp.81-103.

7) R. W. Mottern, R. J. London, and R. A. Morris, "Radiographic Examination of the Shroud of Turin - - - a Preliminary Report," *Materials Evaluation* **38**, 39-44 (1979).

8) V. D. Miller and S. F. Pellicori, "Ultraviolet fluorescence photography of the Shroud of Turin," *Journal of Biological Photography* **49**, 71-85 (1981).

9) A. D. Adler, R. Selzer, and F. DeBlase, in Dorothy Crispino (Ed.), *The Orphaned Manuscript,* Effata Editrice, Via Tre Denti, 1 - 10060 Cantalupa (Torino), Italy (2002) 93-102.

The radiocarbon sample.

1) P. E. Damon, D. J. Donahue, B. H. Gore, A. L. Hatheway, A. J. T. Jull, T. W. Linick, P. J. Sercel, L. J. Toolin, C. R. Bronk, E. T. Hall, R. E. M. Hedges, R. Housley, I. A. Law, C. Perry, G. Bonani, S. Trumbore, W. Woefli, J. C. Ambers, S. G. E. Bowman, M. N. Leese, and M. S. Tite, "Radiocarbon dating of the Shroud of Turin," *Nature,* **337**:611-615 (1989). (Available at http://www.shroud.com/nature.htm)

2) M. S. Benford and J. G. Marino: http://www.shroud.com/pdfs/textevid.pdf

3) M. S. Benford and J. G. Marino: http://www.shroud.com/pdfs/histsupt.pdf

4) R. N. Rogers, "Supportive comments on the Benford-Marino '16th century repairs' hypothesis," *British Society for the Turin Shroud, Shroud Newsletter* **54**, 28-33 (2001).

5) http://www.shroud.com/pdfs/rogers2.pdf.

Chemical tests showed that there is no protein painting medium or protein-containing coating in image areas.

1) E. J. Jumper, A. D. Adler, J. P. Jackson, S. F. Pellicori, J. H. Heller, and J. R. Druzik, "A comprehensive examination of the various stains and images on the Shroud of Turin," *ACS Advances in Chemistry, Archaeological Chemistry* **III:205,** 447-476 (1984).

2) L. A. Schwalbe and R. N. Rogers, "Physics and Chemistry of the Shroud of Turin, a Summary of the 1978 Investigations," *Analytica Chimica Acta* 135 (1982).

3) J. H. Heller and A. D. Adler, "A Chemical Investigation of the Shroud of Turin," *Canadian Society of Forensic Science Journal* **14** (1981), pp.81-103.

Bioplastic polymer claim.

1) H. E. Gove, S. J. Mattingly, A. R. David, and L. A. Garza-Valdes, "A problematic source of organic contamination of linen," *Nuclear Instruments and Methods in Physics Research* **B, 123**, 504-507 (1997).

2) S. J. Mattingly, "The Role of Human Skin Bacteria in the Formation of Photographic-like Images on Linen with Implications for Shroud of Turin Conservation and a Method to Clean Shroud Linen of Microbial Contamination for Reliable Radiocarbon

dating," *British Society for the Turin Shroud, Shroud Newsletter* **54**, November 2001, pp. 12-17.

The image of the dorsal side of the body shows the same color density and distribution as the ventral, and it does not penetrate the cloth any more deeply than the image of the ventral side of the body.
E. J. Jumper, A. D. Adler, J. P. Jackson, S. F. Pellicori, J. H. Heller, and J. R. Druzik, "A comprehensive examination of the various stains and images on the Shroud of Turin," *ACS Advances in Chemistry, Archaeological Chemistry* **III:205,** 447-476 (1984).

The image does not fluoresce under ultraviolet illumination.
V. D. Miller and S. F. Pellicori, "Ultraviolet fluorescence photography of the Shroud of Turin," *Journal of Biological Photography* **49**, 71-85 (1981).

No image formed under the blood stains.
J. H. Heller and A. D. Adler, "A Chemical Investigation of the Shroud of Turin," *Canadian Society of Forensic Science Journal* **14** (1981), pp.81-103.

The image-formation mechanism did not damage the blood.
E. J. Jumper, A. D. Adler, J. P. Jackson, S. F. Pellicori, J. H. Heller, and J. R. Druzik, "A comprehensive examination of the various stains and images on the Shroud of Turin," *ACS Advances in Chemistry, Archaeological Chemistry* **III:205,** 447-476 (1984).

The only image color visible on the back side of the cloth is in the region of the hair.
Giuseppi Ghiberti, "Sindone le immagini 2002," Opera Diocesana Preservazione Fede - Buona Stampa, Corso Matteotti, II, 10121 Torino, Italia, plate 12 and text.

Properties of energetic radiation.
http://physics.nist.gov/PhysRefData/Star/Text/contents.html

History of the technology of linen production.
1) Theophrastus (371 - 287 BC), *Historia Plantarum*, English translation by Sir Arthur Hort, William Heinemann, Ltd., London, 1961, Two volumes.
2) Pliny the Elder, *Natural History*, Book 19.3.16-18 (ca. AD 77).
3) B. Hochberg in *Handspinner's Handbook*, K. Edgerton, L. Knott, Eds. (Windham Center, CT, 1980).

Maillard browning reactions.
1) A. Arnoldi, "Thermal processing and foods quality: analysis and control," in P. Richardson, *Thermal Technologies in Food Processing,* 138-159 (Cambridge UK, Woodhead Publishing, 2001).
2) B. Cämmerer and L. W. Kroh, "Investigation of the influence of reaction conditions on the elementary composition of melanoidins," *Food Chem.* **53,** 55-59 (1995).
3) T. Hoffman, "Characterization of the most intensely colored compounds from Maillard reactions of pentoses by application of color dilution analysis," *Carbohydr. Res.* **313(3-**

4), 203-213 (1998).

Vaporographic hypothesis of image formation.
1) P. Vignon, *The Shroud of Christ* (1902), reprinted by University Books (1970).
2) S. F. Pellicori, "Spectral properties of the Shroud of Turin," *Applied Optics*, **19** (1980), pp. 1913-1920.

The Raes sample.
G. Raes, Rapport d'Analise. La S. Sindone. *Rivista Diocesana Torinese*, 79-83 (1976).

Thermal radiation properties.
G. G. Gubareff, J. E. Janssen, and R. H. Torborg, *Thermal Radiation Properties Survey*, Honeywell Research Center, Minneapolis-Honeywell Regulator Co., Minneapolis, MN, 1960.

Autocatalysis.
1) Feigl, F. and Anger, V., *Spot Tests in Organic Analysis,* Elsevier Pub. Co., New York, 1966.
2) Rogers, R. N., "Differential Scanning Calorimetric Determination of Kinetics Constants of Systems that Melt with Decomposition," Thermochimica Acta **3**, 437 (1972).
3) Rogers, R. N. and Smith, L. C., 1970, "Application of Combined Pyrolysis - TLC to the Study of Chemical Kinetics," J. Chromatog. **48**, 268.

Decomposition of a human body.
Vass, A. A., et al., "Decomposition Chemistry of Human Remains: A New Methodology for Determining the Postmortem Interval," J. Forensic Sci. 2002;47(3):542-553.

INDEX

www.ingramcontent.com/pod-product-compliance
Lightning Source LLC
Chambersburg PA
CBHW041713210326
41598CB00007B/637